McFarlin Library
WITHDRAWN

DATE DUE			
DEC 2 2 1989			

GEOTHERMAL SCALING AND CORROSION

Symposia
presented at
New Orleans, La., 19-20 Feb. 1979,
and Honolulu, Hawaii, 4-5 April 1979

ASTM SPECIAL TECHNICAL PUBLICATION 717
L. A. Casper and T. R. Pinchback,
E G & G Idaho, Inc., Idaho National
Engineering Laboratory, editors

ASTM Publication Code Number (PCN)
04-717000-27

AMERICAN SOCIETY FOR TESTING AND MATERIALS
1916 Race Street, Philadelphia, Pa. 19103

Copyright © by AMERICAN SOCIETY FOR TESTING AND MATERIALS 1980
Library of Congress Catalog Card Number: 80-66077

Note
The Society is not responsible, as a body,
for the statements and opinions
advanced in this publication.

Printed in Baltimore, Md.
December 1980

Foreword

The papers in this volume were presented at two symposia sponsored by the American Society for Testing and Materials through its Committee G-1 on Corrosion of Metals and Subcommittee G01.09 on Corrosion in Natural Waters.

The symposium on Corrosion in Geothermal Systems was cosponsored by the Metallurgical Society of the American Institute of Mining, Metallurgical, and Petroleum Engineers. This symposium was held in New Orleans, La., on 19-20 Feb. 1979.

The symposium on Geothermal Scaling and Corrosion was cosponsored by the Industrial and Engineering Chemistry Division of the American Chemical Society and was held on 4-5 April 1979 in Honolulu, Hawaii.

Both symposia were cochaired by L. A. Casper and T. R. Pinchback, both of the Idaho National Engineering Laboratory, E G & G Idaho, Inc. These men also served as editors of this publication.

Related
ASTM Publications

Corrosion in Natural Environments, STP 558 (1974), $29.75, 04-558000-27

MiCon 78: Optimization of Processing, Properties, and Service Performance Through Microstructural Control, STP 672 (1979), $59.50, 04-672000-28

A Note of Appreciation to Reviewers

This publication is made possible by the authors and, also, the unheralded efforts of the reviewers. This body of technical experts whose dedication, sacrifice of time and effort, and collective wisdom in reviewing the papers must be acknowledged. The quality level of ASTM publications is a direct function of their respected opinions. On behalf of ASTM we acknowledge with appreciation their contribution.

ASTM Committee on Publications

Editorial Staff

Jane B. Wheeler, *Managing Editor*
Helen M. Hoersch, *Associate Editor*
Helen P. Mahy, *Senior Assistant Editor*
Allan S. Kleinberg, *Assistant Editor*

Contents

Introduction   1

Chemistry and Materials in Geothermal Systems—R. L. MILLER   3

Thermodynamics of Corrosion for Geothermal Systems—D. D. MACDONALD   10

Material Selection Guidelines for Geothermal Power Systems—An Overview—MARSHALL CONOVER, PETER ELLIS, AND ANNE CURZON   24

Application of Linear Polarization Techniques to the Measurement of Corrosion Rates in Simulated Geothermal Brines—M. J. DANIELSON   41

Corrosion Protection of Solar-Collector Heat Exchangers and Geothermal Systems by Electrodeposited Organic Films—G. H. SCHNAPER, V. R. KOCH, AND S. B. BRUMMER   57

Preliminary Evaluation of Materials for Fluidized Bed Technology in Geothermal Wells at Raft River, Idaho, and East Mesa, California—W. J. DIRK, C. A. ALLEN, AND R. E. McATEE   69

Surface Corrosion of Metals in Geothermal Fluids at Broadlands, New Zealand—W. R. BRAITHWAITE AND K. A. LICHTI   81

Corrosion in Geothermal Brines of the Salton Sea Known Geothermal Resource Area—S. D. CRAMER AND J. P. CARTER   113

Corrosion of Structural Steels in High-Salinity Geothermal Brine—W. T. LEE AND D. KRAMER   142

Degradation of Elastomers in Geothermal Environments—C. ARNOLD, JR., K. W. BIEG, AND J. A. COQUAT   155

Polymeric and Composite Materials for Use in Systems Utilizing Hot, Flowing Geothermal Brine III—L. E. LORENSEN, C. M. WALKUP, AND C. O. PRUNEDA   164

Concrete Polymer Materials as Alternative Materials of Construction for Geothermal Applications—Field Test Evaluations—J. J. FONTANA AND A. N. ZELDIN 180

Organosiloxane Polymer Concrete for Geothermal Environments—A. N. ZELDIN, L. E. KUKACKA, J. J. FONTANA, AND N. R. CARCIELLO 194

Treatment Methods for Geothermal Brines—S. L. PHILLIPS, A. K. MATHUR, AND WARREN GARRISON 207

Chemical Logging of Geothermal Wells—R. E. McATEE, C. A. ALLEN, AND L. C. LEWIS 225

Round-Robin Evaluation of Methods for Analysis of Geothermal Brine—J. C. WATSON 236

Summary 259

Index 261

Introduction

Geothermal energy is one of many technologies being developed to meet critical needs for heat and power. Geothermal sources have been utilized in isolated instances for many years, primarily as a means of providing local heating. Recent efforts have been directed toward greatly increasing the electrical generating capacity of geothermal systems and, to a lesser extent, the process and space heat produced.

Corrosion and in some cases scaling have presented problems in many geothermal systems. Dissolved material in geothermal waters can exhibit aggressive corrosion properties or have the tendency to deposit large amounts of mineral scale Either property can seriously shorten the service life of piping in the source well, the process plant, or the reinjection well.

Scaling and corrosion constitute serious technical barriers to the utilization of geothermal resources. Because of the large quantities of water that must be processed to obtain heat, many conventional approaches to these problems, such as the use of inhibitors, are not economically viable. These problems can be controlled through innovative applications of materials science and chemistry.

The papers in this special technical publication should be of interest to all those who deal with materials problems in geothermal systems. Such materials problems are approached from several points of view in this collection of papers, including fundamental scientific investigations, field studies of materials in geothermal systems and some new alternative materials, and some aspects of the chemistry of the geothermal fluids. This should provide a useful reference for both the scientist/engineer who must deal with specific geothermal systems and those in management/operations who require an overview of the technology of materials problems.

The two symposia represented in this book were organized to provide a forum within the materials science and chemistry communities for the presentation and discussion of current research into the problem. Appreciation is expressed to the Metallurgical Society of the American Institute of Mining, Metallurgical, and Petroleum Engineers and to the American Chemical Society for their joint cooperation with ASTM in these symposia.

L. A. Casper
T. R. Pinchback
E G & G Idaho, Inc., Idaho National Engineering Laboratory, Idaho Falls, Idaho 83401; symposium cochairmen and editors.

R. L. Miller[1]

Chemistry and Materials in Geothermal Systems

REFERENCE: Miller, R. L., "**Chemistry and Materials in Geothermal Systems,**" *Geothermal Scaling and Corrosion, ASTM STP 717*, L. A. Casper and T. R. Pinchback, Eds., American Society for Testing and Materials, 1980, pp. 3-9.

ABSTRACT: The development of a geothermal fluid is traced, from its origin as meteoric water precipitating on the earth's surface, as it flows through the soils and rocks of geological formations, to the point where it returns to the surface as a hot spring, geyser, well, or other form. Water of magmatic origin is also included. The tendency of these hydrothermal fluids to form scale by precipitation of a portion of their dissolved solids is noted. A discussion is presented of types of information required for materials selection for energy systems utilizing geothermal fluids, including pH, temperature, the speciation of the particular geothermal fluid (especially its chloride, sulfide, and carbon dioxide content), and various types of corrosive attack on common materials. Specific examples of responses of materials to geothermal fluids are given.

KEY WORDS: corrosion, geothermal environment, materials selection, scaling

Corrosion and scaling in geothermal energy recovery systems are two of the more important problems that require the close attention of chemists, geologists, and materials scientists. These problems result from the nature of the geothermal fluids encountered as systems are designed and built to extract energy from these resources. The geothermal fluid, in turn, is a result of the environment from which it is extracted. The history of the fluid as it comes into contact with various minerals is the key to understanding the tendency of these fluids to promote scaling and corrosion and, as a consequence, making rational materials selections for plant construction.

Origin and Chemical Characteristics of Geothermal Fluids

Geothermal resources vary in character and distribution, but some generalities are evident. First, there is a close relationship between earth-

---
[1] Senior scientist, E G & G Idaho, Inc., Idaho Falls, Idaho 83401.

quake and volcano belts and geothermal fields. Thus, the region around the Pacific Ocean is especially important in regard to known geothermal resource areas. These areas include such well-known fields as those in Japan, New Zealand, Central America, and the western regions of the United States, including Alaska and Hawaii. A second generality is that the salinity of the geothermal water available for development increases with the resource temperatures.

The two geothermal fluids available are steam and water. Steam-dominated resources occur less frequently than water-dominated systems. Because of the greater abundance of water-dominated systems, and because my own experience lies in this area, these resources will be emphasized in this paper. Hot dry rock and geopressured systems are mentioned only in passing.

The water in a geothermal resource is of several types, as noted by Ellis and Mahon [1].[2] Meteoric and magmatic waters provide the primary sources and seem to be the most abundant. Meteoric water results from rain and snowfall and differs from standard normal ocean water in its content of deuterium and heavy oxygen (^{18}O). Meteoric water generally has lower ^{18}O content than waters originating from magmatic sources or meteoric water that has been in contact with siliceous materials for long periods. The evidence to date suggests that geothermal fluids contain both meteoric and magmatic water; the relative amounts of each are characteristic of a particular resource.

Consider now a drop of water falling through the atmosphere. The dissolved gases in this unit of meteoric water will be in equilibrium with the atmosphere as the water strikes the earth. The important chemical agents in this fluid are water, oxygen, and carbon dioxide; all are powerful weathering agents. Water is both an active mineral solvent and a carrier for the dissolved minerals. Oxygen acts on sulfides and other minerals to form soluble products; it also acts on dead plant and animal matter to increase the rate of decay. Carbon dioxide is a mild oxidizing agent, but its primary function is found in its action on the kinds and amounts of carbonate species.

Returning briefly to plant and animal decay products, one finds that certain of these organic substances are strong weathering agents. In particular, the humic and fulvic acid fractions are potent in bringing minerals into solution and in forming relatively stable coordination compounds. Humic acids are characterized by carboxylic acid and phenolic functional groups. Singer and Navrot [2] found that a significant portion of the total metal content of metal-rich basalts was leached in a few hundred hours of contact with humic acid solutions at temperatures as low as 323 K.

Within the near-surface environment, the principal chemical weathering

---

[2]The italic numbers in brackets refer to the list of references appended to this paper.

agents are water, oxygen, carbon dioxide, and decay products from plant and animal life. As these agents are transported deeper into the crust by a process termed elutriation, they bring about chemical changes and increase the load of dissolved solids and gases in the water generally.

As the water droplet penetrates still deeper into the crust, the temperature of the surrounding rock increases because of a heat flux from the molten core of the earth toward the cooler surface. This heat flux averages 0.063 J/m^2 and results in a thermal gradient that varies with the thermal conductivity of the area. Its value is usually about 20 K/km. In geothermally active areas, the gradients are much higher; for example, in southern Idaho the gradient is of the order of 100 K/km. This heat flux results in heating of the meteoric water and in accelerating rates of reaction between water and rock.

Magmatic water is released from solidifying magma. The heat source of the magma is the decay of radioactive elements, with uranium, thorium, and potassium providing most of the heat [3]. As the radioactive elements are consumed, less decay heat is produced, and some of the magma crystallizes. There are local hot spots, as indicated by volcanic action, but for present purposes these can be ignored. A portion of the solidifying magma contains chemically bound water. As the magma solidifies, the water is released from the rock and enters the geological formation as magmatic water. In water-dominated fields this magmatic water eventually cools enough to form a condensed phase. (In a steam-dominated field a substantial portion of the geothermal fluid will remain as a vapor.) Although the liquid water will be free of oxygen and organic matter, it will be saturated with dissolved rock matter and contain gases such as carbon dioxide and hydrogen sulfide. The former is a product of carbonate decrepitation; the latter results from the decomposition of sulfide minerals.

The hot meteoric water and the hot magmatic water will react with other rocks in the geological formation. As they pass from one stratum to another, some minerals will be deposited and others dissolved. This geothermal fluid varies in composition according to local lithology, so that geothermal fields show profound differences in water chemistry. Furthermore, there are variations from well to well within a field, and any single well may vary in water composition with time and with the rate of water production.

Typically, chloride and sulfate are the most abundant anions, whereas sodium and calcium are the most abundant cations. Silica is the most abundant nonionic species in most geothermal waters. The ionic species are formed by the passing of water through beds of evaporites and by the decomposition of various minerals. The silica results from dissolution of a number of mineral species, including quartz. The ionic strength of most geothermal fluids is so high that serious deviations from ideal models, such as the Debye-Huckel theory, have been observed.

The hot "rock soup" is the fluid that must be worked with. This is the fluid from which heat is abstracted for electrical power production, for space heating, or for process heat in a number of applications. The papers in this volume address the character of this rock soup and materials for containing, transmitting, and handling it.

Problems in Use

Once the geothermal fluid is available for use, two important technological problems confront us; these are scaling and corrosion. The scaling minerals are typically silica and calcite, although sulfides are important in some fields. Corrosion products are a second source of scale. Both of these sources of scale are important because of the adverse effect of scale on heat transfer and pumping efficiencies.

The species in the water that are of greatest interest in relation to corrosion are hydrogen ion, chloride ion, hydrogen sulfide, carbon dioxide, ammonia, and sulfate ion [4]. Some generalizations related to the corrosive effects of these species are noted here:

Hydrogen Ion—The corrosion rates of most materials increase as the pH of the fluid decreases. This trend is especially evident in plain carbon steels. The susceptibility of stainless steels to stress corrosion cracking increases with increasing hydrogen ion content.

Chloride—The chloride ion causes breakdown of the passive films that provide some protection to the substrate metals. This frequently happens in localized areas and results in pitting and other forms of localized corrosion, as well as uniform corrosion. Chlorides also form relatively stable complex ions or coordination compounds that can result in accelerated corrosion [5].

Hydrogen Sulfide—Copper and its alloys are attacked by hydrogen sulfide. Copper alloys with nickel additions are attacked at rates that are several times those of similar, but nickel-free, alloys. Sulfide stress cracking in high-strength steels is a potential problem that should be recognized in geothermal systems.

Carbon Dioxide—Carbon dioxide is a mild oxidizing agent that causes increased corrosion of plain carbon steels. However, the primary effects of carbon dioxide in geothermal systems are on carbonate speciation [6] and pH.

Ammonia—This species frequently leads to increased corrosion of copper-based alloys and is especially important in relation to stress corrosion cracking. Mild steels are also adversely affected by ammonia.

Sulfate—Sulfate is the primary aggressive ion in some geothermal fluids. It is not, however, as harsh an ion in this role as the chloride ion [7].

Oxygen is usually present in such low concentrations in most geothermal fluids that it can be neglected. On the other hand, the inadvertent intrusion of even traces of this gas into hot geothermal fluids has led to greatly accelerated corrosion. One case in point involves a gravity-fed space-heating system that contained a nominal 15 μg/litre of oxygen; when the system was expanded, and pumps were used to pressurize the water, the oxygen content rose to 65 μg/litre, and the corrosion rate in the new system was several times that in the older part of the installation. The combination of oxygen and chloride is especially bad and has led to catastrophic failure due to stress corrosion cracking in at least one geothermal facility using an austenitic stainless steel in a critical application.

Materials for Geothermal Service

The selection of materials for applications in geothermal environments is one area in which chemists, metallurgists, and materials scientists can make a significant contribution to the geothermal effort. The responses of plain carbon and low-alloy steels, stainless steels, nickel-base alloys, copper-base alloys, aluminum, titanium, and other materials are briefly reviewed in relation to unflashed fluids, such as those at Raft River, Idaho, and East Mesa, Calif.

Plain Carbon and Low-Alloy Steels—Experience with these materials indicates that severe pitting and crevice corrosion occur, in addition to less severe uniform corrosion. The alloyed materials show more resistance but should not be selected without a site-specific test that simulates the intended usage of the materials. These materials are subject to hydrogen blistering, especially in sulfide environments, and to sulfide stress cracking in the higher strength grades. Galvanic corrosion is strongly dependent on the coupling material, and, in my experience, the corrosion of low-carbon steels is greater when the cathodic material is brass rather than stainless steel.

Stainless Steels—Ferritic alloys are subject to pitting and crevice corrosion, especially in stagnant conditions. However, some high-alloy ferritic stainless steels have shown very good resistance to a moderate-temperature geothermal fluid [8]. Austenitic alloys are prone to stress corrosion cracking, and at least one instance of this form of attack has been seen in geothermal usage. These alloys are also subject to localized and intergranular corrosion. Testing of these alloys should include the evaluation of oxygen intrusion into the geothermal fluid. Martensitic alloys suffer from the same forms of corrosion as ferritic and austenitic stainless steels. Stainless steels should be tested in the proposed environment prior to being selected for use. Further, the testing should simulate the effects of oxygen on the alloys to evaluate their response to a stress corrosion cracking environment.

Nickel-Base Alloys—These alloys are generally considered to be the superalloys and are typically very resistant to corrosion. However, they are subject to localized corrosion, such as pitting and crevice corrosion, especially in high-temperature chloride environments.

Copper-Base Alloys—These alloys are subject to attack by hydrogen sulfide, and care must be exercised in their selection. The nickel-free alloys are more resistant than nickel-containing alloys of similar copper content. Since copper alloys are subject to stress corrosion cracking in ammonia-containing waters, an analysis for ammonia should be performed as part of the geochemical surveys of a field. Waters containing more than a few milligrams of ammonia per litre should be tested for their aggressiveness toward copper alloys.

Aluminum Alloys—Aluminum alloys have been recommended for desalination applications. Their use in geothermal environments cannot be recommended because of the very high corrosion rates that have been encountered.

Titanium Alloys—Titanium and its alloys are among the most resistant materials tested to date. However, experience in other environments suggests that at temperatures above 573 K these materials are subject to crevice corrosion. Fluorides specifically attack titanium, and, therefore, this material should be avoided where fluoride-rich waters are encountered.

Other Alloys—Cobalt-base alloys are noted for their hardness and can be used in geothermal environments when this characteristic is desirable. Zirconium has very good corrosion resistance in a number of environments, but the cost of these materials is such that no applications to geothermal environments are projected.

Conclusion

The corrosion and scaling problems encountered in geothermal systems are a direct result of geochemical changes that occur in the water. Until sufficient data are available to permit materials selection on the basis of water chemistry and other measurable fluid properties, materials selection for geothermal environments should be supported by site-specific testing that simulates those conditions under which the materials will be used. Eventually, sufficient data on and insight into the chemistry of geothermal fluids will be available to permit adequate materials selection without long-term corrosion tests.

Acknowledgment

This work was supported by the U.S. Department of Energy Assistant Secretary for Resource Applications, Office of Geothermal Energy, and the report was authored by a contractor of the U.S. government, under DOE Contract No. DE-AC07-76ID01570.

References

[1] Ellis, A. J. and Mahon, W. A. J., *Chemistry and Geothermal Systems*, Academic Press, New York, 1977, p. 28.
[2] Singer, A. and Navrot, J., *Nature*, Vol. 262, 1962, pp. 479-481.
[3] Bullard, E., *Geothermal Energy, Earth Sciences*, Vol. 12, United Nations Educational, Scientific and Cultural Organization, New York, 1973, pp. 19-28.
[4] DeBerry, D. W., Ellis, P. F., and Thomas, C. C., *Materials Selection Guidelines for Geothermal Power Systems*, USDOE Report ALO/3903-1, U.S. Department of Energy, Washington, D.C., Sept., 1978.
[5] Foley, R. T., *Journal of the Electrochemical Society*, Vol. 122, 1975, pp. 1493-1494.
[6] Garrels, R. M. and Christ, C. L., *Solutions, Minerals, and Equilibria*, Harper and Row, New York, 1965, pp. 74-92.
[7] Suciu, D. F. and Miller, R. L., *Short-Term Pilot Cooling Tower Tests*, USDOE Report EGG-FM-5087, U.S. Department of Energy, Washington, D.C., Jan. 1980.
[8] Miller, R. L., *Results of Short-Term Corrosion Evaluation Tests at Raft River*, USDOE Report TREE-1176, U.S. Department of Energy, Washington, D.C., Oct. 1977.

D. D. Macdonald[1]

Thermodynamics of Corrosion for Geothermal Systems

REFERENCE: Macdonald, D. D., "**Thermodynamics of Corrosion for Geothermal Systems,**" *Geothermal Scaling and Corrosion, ASTM STP 717,* L. A. Casper and T. R. Pinchback, Eds., American Society for Testing and Materials, 1980, pp. 10–23.

ABSTRACT: Potential-pH diagrams, which summarize the thermodynamic properties of a system, are extensively used for interpreting the corrosion behavior of metals in condensed media. In this paper, E-pH diagrams are presented for iron and nickel in sulfide-containing high-salinity geothermal brine at 25°C and 250°C. The diagrams have been constructed on the assumption of a constant total sulfide constraint which leads to nonlinearity of the equilibrium relationships in potential-pH space. The application of these diagrams in the interpretation of chemical and corrosion processes in geothermal systems is briefly discussed.

KEY WORDS: geothermal brine, iron, nickel, thermodynamics, potential-pH diagrams, corrosion, scaling

In the analysis of corrosion phenomena, it is frequently necessary to have some knowledge of the equilibrium properties of the system. This information not only indicates whether or not any given process is spontaneous, but also defines the conditions that must be achieved in order to minimize the effects of corrosion in practical systems. Thus, cathodic protection requires that the electrochemical potential of the structure be displaced (electronically or by coupling to a sacrificial anode) into the thermodynamically immune region of potential-pH space. In this state, continued corrosion is thermodynamically *not* possible, so that the degree of protection is independent of the kinetics of the interfacial processes. On the other hand, anodic protection requires maintenance of the potential in the so-called "passive" region of potential-pH space, in which a stable or metastable corrosion product film is maintained at the metal surface. Since the overall corrosion process is spontaneous, protection of the metal structure is a kinetic rather than a thermodynamic property of the system.

[1]Professor, Metallurgical Engineering, Ohio State University, Columbus, Ohio 43210.

In this paper, methods for estimating thermodynamic properties of metal/water (brine) systems under geothermal conditions are discussed. The use of these data for the derivation of potential-pH diagrams and the application of the diagrams to the analysis of corrosion phenomena in geothermal systems is described. Particular reference is made to the equilibrium properties of iron and nickel in high-salinity brine, since these metals are major components of alloys that are now being used in the fabrication of production equipment. Potential-pH diagrams for chromium and titanium under similar conditions are given elsewhere [1].[2]

Theory

Calculation of Gibbs Energy Changes of Reactions from Isothermal Free Energies of Formation

The conventional method of evaluating the Gibbs energy change for a reaction at a temperature, T, involves the use of Eq 1

$$\Delta G_T^0 = \Sigma_P \nu_P \Delta_f G_T^0 - \Sigma_R \nu_R \Delta_f G_T^0 \tag{1}$$

where $\Delta_f G_T^0$ is the Gibbs energy formation for each component in the reaction, and ν_P and ν_R are the stoichiometric coefficients for the products and reactants, respectively.

The standard Gibbs energy of a substance at temperature T, G_T^0, can be expressed in terms of the standard entropy and standard Gibbs energy at T_1, the reference temperature (defined to be 298 K), and the heat capacity of the system over the temperature interval T_1 to T by Eq 2

$$G_T^0 = G_{298}^0 - S_{298}^0 (T - 298) - T \int_{298}^{T} \frac{C_P}{T} dT + \int_{298}^{T} C_P^0 dT \tag{2}$$

The conventional standard Gibbs energy of formation at temperature T, that is, the Gibbs energy change for an isothermal reaction of the component elements to form the product, can then be written as

$$\Delta_f G_T^0 = \Delta_f G_{298}^0 - (T - 298) \Delta_f S_{298}^0 - T \int_{298}^{T} \frac{\Delta_f C_P}{T} dT + \int_{298}^{T} \Delta_f C_P^0 dT \tag{3}$$

where $\Delta_f S_{298}^0$ is the change in standard entropy for the formation reaction. If a phase transition occurs in the range 298 to T K, it is necessary to include a term allowing for the changes in entropy and enthalpy.

[2]The italic numbers in brackets refer to the list of references appended to this paper.

The Gibbs energy change for a particular reaction may then be calculated by substitution into Eq 1 of values for $\Delta_f G_T^0$.

An Alternative Method to Evaluate Free Energy Changes of Reactions

In order to reduce the number of calculations required to evaluate the Gibbs energy change of reaction, Macdonald et al [2,3] proposed a method that involves the use of Eq 4

$$'G_T^{0'} = \Delta_f G_{298}^0 - S_{298}^0 (T - 298) - T \int_{298}^{T} \frac{C_P^0}{T} dT + \int_{298}^{T} C_P^0 \, dT \quad (4)$$

to evaluate quantities, $'G_T^{0'}$, from which values of ΔG_T^0 may be obtained directly through

$$\Delta G_T^0 = \Sigma_P \nu_P \, 'G_T^{0'} - \Sigma_R \nu_R \, 'G_T^{0'} \quad (5)$$

The symbol, $'G_T^{0'}$, by definition, refers to the Gibbs energy of formation of a species at temperature T from the elements at 298.15 K and is obtained by integrating the fundamental differential Eq 6

$$dG = VdP - SdT \quad (6)$$

We have previously proposed [4] to represent the Gibbs energy change, $'G_T^{0'}$, by the notation $\Delta_f G_T^{0*}$.[3]

Equation 4 is now written as

$$\Delta_f G_T^{0*} = \Delta_f G_{298}^0 - S_{298}^0 (T - 298) - T \int_{298}^{T} \frac{C_P^0}{T} dT + \int_{298}^{T} C_P^0 \, dT \quad (7)$$

and the Gibbs energy change for the reaction is then given by Eq 8

$$\Delta G_T^0 = \Sigma_P \nu_P \Delta_f G_T^{0*}(P) - \Sigma_R \nu_R G_T^{0*}(R) \quad (8)$$

---

[3]This notation was chosen in order to have the same form as that commonly used for the change in a thermodynamic function, which occurs on formation of a substance from its elements. In the present case, the notation indicates that the formation reaction is non-isothermal (superscript *) and that the substance formed is at a temperature of T. Clearly $\Delta_f G_{298}^{0*}$ is identical to the conventional isothermal quantity $\Delta_f G_{298}^0$.

The advantage of this technique is that it is not necessary to calculate values of $\Delta_f G_T^0$ for every compound at each temperature. This results in a considerable reduction in the number of calculations that must be performed to describe the equilibrium behavior of a complete metal/water system over a wide range of temperatures.

Evaluation of $\Delta_f G_T^{0*}$

For nondissolved substances, accurate heat capacity functions of the form

$$C_P^0 = A + BT + CT^{-2} \quad (9)$$

are available [5-9], and $\Delta_f G_T^{0*}$ can be calculated directly from Eq 7.

Directly measured heat capacity data for ionic (dissolved) species are generally not available, and, consequently, Gibbs energy changes for each species must be estimated. The calculations are approached more easily by considering the integral containing the temperature-dependent C_P^0 functions in terms of entropy. If the following approximation is used [2]

$$\int_1^T C_P^0 \, dT \approx C_P^0 (T - T_1) = \frac{T - T_1}{\ln(T/T_1)} (S_T^0 - S_{T_1}^0) \quad (10)$$

where S_T^0 and $S_{T_1}^0$ are absolute entropies of the ion at temperature T and T_1, respectively, then Eq 7 for an ionic species may be transformed into Eq 11, with an error of generally less than 1 percent [3].[4]

$$\Delta_f G_T^{0*} = \Delta_f G_{298}^0 - (TS_T^0 - 298 S_{298}^0) + \frac{T - 298}{\ln(T/298)} (S_T^0 - S_{298}^0) \quad (11)$$

The absolute entropies of ions at elevated temperatures can be estimated using the correspondence principle of Criss and Cobble [10]

$$S_T^0 = a + b S_{T_1}^0 \quad (12)$$

where a and b are constants that are unique for a given temperature and class of ion. Entropies of ions based on the conventional scale, where the entropy of the hydrogen ion at 298 K is defined as zero, may be converted to the absolute scale using Eq 13

$$S^0 \text{ (absolute)} = S^0 \text{ (conventional)} - 20.9z \; (JK^{-1}mol^{-1}) \quad (13)$$

where z is the ionic charge.

---

[4]Strictly speaking, Eq 11 should be written in terms of the chemical potentials.

Derivation of Potential-pH Diagrams

The computational methods used for deriving potential-pH relationships for metal/water systems at elevated temperatures closely follow those described previously [2-4,13-17]. Briefly, the equilibria used to describe the chemistry of a system in which sulfur does not undergo a change in oxidation state are written in the form [13,16]

$$(NR)R + (NH2S)H_2S + (NH)H^+ + (NH2)H_2 \\ = (NP)P + (NCl)Cl^- + (NH2O)H_2O \quad (14)$$

where the quantities in parentheses refer to the stoichiometric coefficients, R is a general reactant, and P a general product.[5] In cases that involve the oxidized sulfur species SO_4^{2-} and HSO_4^-, the equilibria are written in the closely related forms [16]

$$(NR)R + (NSO4)SO_4^{2-} + (NH)H^+ + (NH2)H_2 \\ = (NP)P + (NH2O)H_2O \quad (15)$$

$$(NR)R + (NHSO4)HSO_4^- + (NH)H^+ + (NH2)H_2 \\ = (NP)P + (NH2O)H_2O \quad (16)$$

The decomposition of Reaction 14 into the appropriate half-cell reaction gives

$$(NR)R + (NH2S)H_2S + [(NH) + 2(NH2)]H^+ + 2(NH2)e^- \\ = (NP)P + (NCl)Cl^- + (NH2O)H_2O \quad (17)$$

and

$$2(NH2)H^+ + 2(NH2)e^- = (NH2)H_2 \quad (18)$$

Comparable decompositions are easily performed for Reactions 15 and 16 and involve the sulfur oxyanions SO_4^{2-} and HSO_4^-. Both half-cell reactions are written in the reduction sense so as to conform with the Stockholm convention for the analysis of electrochemical cells. Because the potential of Reaction 17 is referred to the standard hydrogen scale (Reaction 18), the potential-pH relationship is written as

$$E = C(1) + C(2) \log a_R + C(3) \log a_{H_2S} + C(4) \log a_P \\ + C(5) \log a_{Cl^-} + C(6) \log a_{H_2O} + C(7)\text{pH} \quad (19)$$

---

[5]The formalism is easily extended to handle more complex reactions—for example, those that involve multiple reactants and products.

where $C(1) = -\Delta G_T^0/zF$;
$C(2) = 2.303RT(NR)/zF$;
$C(3) = 2.303RT(NH2S)/zF$;
$C(4) = -2.303RT(NP)/zF$;
$C(5) = -2.303RT(NCl)/zF$;
$C(6) = -2.303RT(NH2O)/zF$;
$C(7) = -2.303RTx/zF$;
$x = (NH) + 2(NH2)$; and
$z = 2(NH2)$.

Likewise, the potential-pH relationships for equilibria involving sulfur oxyanions are written as

$$E = C'(1) + C'(2) \log a_R + C'(3) \log (a_{HSO_4^-} \text{ or } a_{SO_4^{2-}}) + C'(4) \log a_P + C'(6) \log a_{H_2O} + C'(7)pH \qquad (20)$$

where the coefficients $C'(1)$ to $C'(7)$ are defined in a manner analogous to those for Eq 19.

From Eqs 19 and 20, it is clear that derivation of the potential-pH relationships requires a knowledge of the change in standard Gibbs energy (ΔG_T^0) for the cell Reactions 14 to 16 (see Eq 8) and of the activities of water (H_2O), Cl^-, hydrogen sulfide (H_2S), HSO_4^-, SO_4^{2-}, R, and P. In the work discussed in this paper, the activity of water is assumed to equal 1 (Raoult's law standard state); the activity of H_2S is derived from the constant total sulfide constraint (see Activity Functions, which follows; a_{Cl^-} is calculated from the system composition; and the activity of HSO_4^-, SO_4^{2-}, R, and P are assigned arbitrary values.

Activity Functions

Because the total sulfide content of a geothermal brine is fixed by geochemical processes, we considered [16] the conventional method of fixing the activities of individual sulfide species (H_2S, HS^-, S^{2-}), irrespective of the pH of the medium, to be inappropriate. Instead, we believe that a more realistic description of a geothermal system involves calculation of the activity of dissolved H_2S in the system as a function of pH, and then substitution of this quantity into Eq 19 to derive the potential-pH relationship for the reaction of interest.

The equilibria used to describe the chemistry of the H_2S/H_2O system are written as

$$H_2S = H^+ + HS^- \quad (K_1) \qquad (21)$$

$$HS^- = H^+ + S^{2-} \quad (K_2) \qquad (22)$$

where K_1 and K_2 are, respectively, the first and second dissociation constants [5] for H_2S. From the definitions of K_1 and K_2, we obtain the activity of H_2S in terms of the total sulfide concentration [$(H_2S) + (HS^-) + (S^{2-}) = (S)$] as follows

$$a_{H_2S} = \gamma_1\gamma_2 a_{H^+}^2 [S] / \{\gamma_1\gamma_2 a_{H^+}^2 + \gamma_2 K_1 a_{H^+} + \gamma_1 K_1 K_2\} \quad (23)$$

where γ_1 and γ_2 are activity coefficients for univalent and divalent ions, respectively. The activity coefficients for the ions in the high-salinity brine (ionic strength = 3.0 m) were estimated by using the extended Debye-Huckel expressions given by Naumov et al [5]. The activity coefficients so calculated are best regarded as rough estimates only. Comparing γ_1 with $\gamma_\pm$ for sodium chloride (NaCl) at temperatures as high as 250°C indicates that γ_1 may be in error by as much as 16 percent. However, simple calculation [16] indicates that the error in the equilibrium potential from this source is less than 0.007 V; this error is negligible compared with the range of the potential of interest (>3 V).

The substitution of Eq 13 into Eq 6 therefore gives

$$E = C(1) + C(2) \log a_R + C(4) \log a_p + C(5) \log a_{Cl^-}$$
$$+ C(6) \log a_{H_2O} + C(3) \log [S] + [C(7) - 2C(3)] \text{pH}$$
$$- C(3) \log \{a_{H^{2+}} \gamma_1\gamma_2 + \gamma_2 K_1 a_{H^+} + \gamma_1 K_1 K_2\} \quad (24)$$

According to Eq 24, the potential varies nonlinearly with the pH, in contrast with the linear relationship predicted by Eq 19 when constant activities are assumed for all species involved in the half-cell reactions.

Other activity functions can be defined for the derivation of potential-pH diagrams for complex systems. For example, Macdonald and Hyne [13] used a constant H_2S pressure constraint in their thermodynamic analyses of the iron/H_2S/H_2O system at elevated temperatures.

Thermodynamic Data

Because of the many reactions used to describe the thermodynamic behavior of iron and nickel in sulfide-containing geothermal brine at elevated temperatures, it has not proved feasible to list all the equilibria considered or the thermodynamic data used in this paper. The data employed are listed in Ref 17, and copies of the input data and reaction statements can be obtained from the authors. Briefly, the thermodynamic data were taken from the compilation of Naumov et al [5]. Their data generally agree with those in other compilations, such as the U.S. Bureau of Mines monographs [8,9] and publications of the National Bureau of Standards [6,7].

Discussion

Potential-pH diagrams for iron and nickel in high-salinity geothermal brine containing 10 ppm of total dissolved sulfide ($H_2S + HS^- + S^{2-}$) are shown in Figs. 1 to 4. Also plotted are the corresponding diagrams for the sulfur/water system, whose equilibrium relationships are designated by the light broken lines and by numbers ending with the letter "s" (1s, 2s,

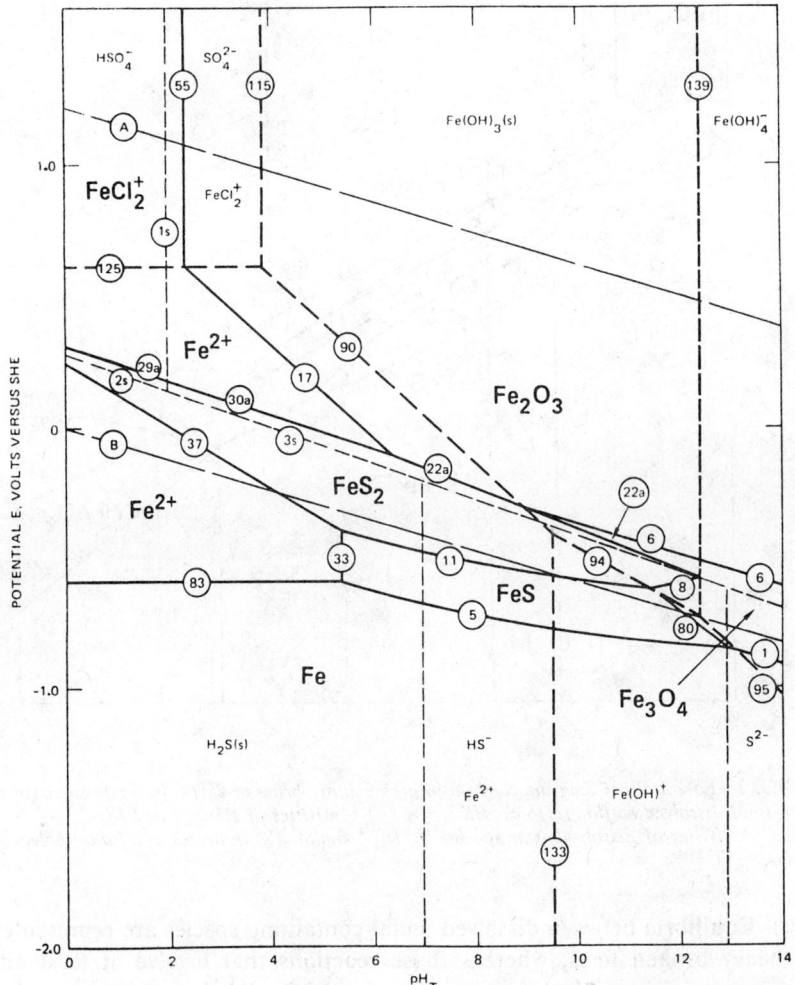

FIG. 1—*Potential-pH diagram for iron in high-salinity brine at 25°C in the presence of 10 ppm total dissolved sulfide ($H_2S + HS^- + S^{2-}$). Activities of HSO_4^- and $SO_4^{2-} = 10^{-6}$ molal. Activities of dissolved iron species = 10^{-4} molal. (S) indicates a soluble molecular species.*

18   GEOTHERMAL SCALING AND CORROSION

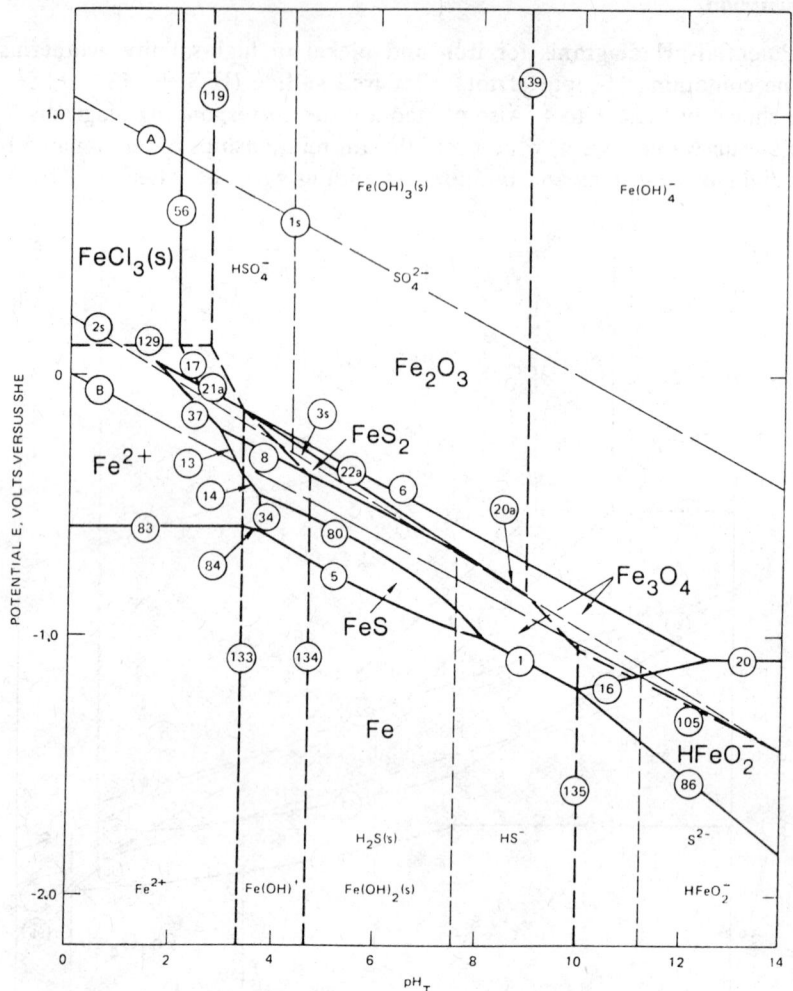

FIG. 2—*Potential-pH diagram for iron in high-salinity brine at 250°C in the presence of 10 ppm total dissolved sulfide ($H_2S + HS^- + S^{2-}$). Activities of HSO_4^- and $SO_4^{2-} = 10^{-6}$ molal. Activities of dissolved iron species $= 10^{-4}$ molal. (S) indicates a soluble molecular species.*

etc.). Equilibria between dissolved metal-containing species are represented by heavy broken lines, whereas those reactions that involve at least one metal-containing solid phase are represented by solid lines. Lines A and B represent the thermodynamic equilibrium limits for the stability of liquid water. For example, at voltages above line A, oxygen spontaneously evolves, whereas at voltages more negative than those given by line B, hydrogen evolution tends to occur. All four diagrams were derived for constant

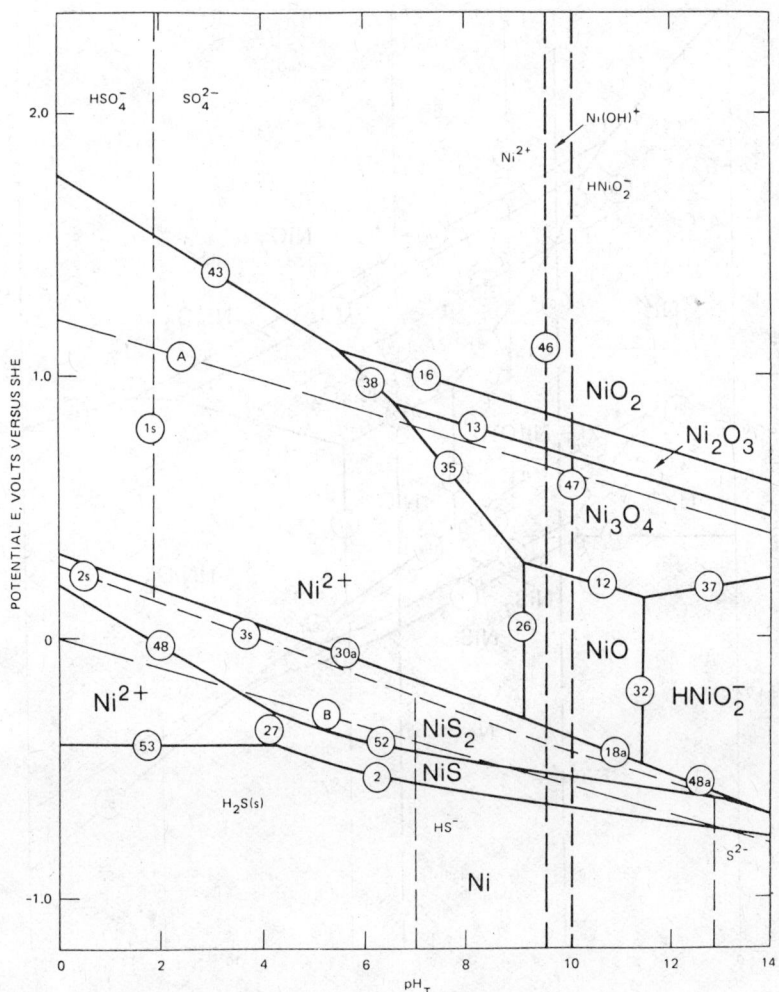

FIG. 3—*Potential-pH diagram for nickel in high-salinity brine at 25°C in the presence of 10 ppm total dissolved sulfide (H_2S + HS^- + S^{2-}). Activities of HSO_4^-, SO_4^{2-}, and dissolved nickel species = 10^{-6} molal. (S) indicates a soluble molecular species.*

activities for HSO_4^- and SO_4^{2-} equal to 10^{-6} m. In the case of the iron/brine system, the activities of dissolved metal-containing species were arbitrarily set equal to 10^{-4} m, whereas for the nickel/brine system dissolved species activities of 10^{-6} m were assumed. The higher activities for the iron/brine system were required so as to exceed the calculated minimum solubilities of the oxide phases at elevated temperatures. Had the lower activity been assumed, the oxides would not have appeared as stable phases on the diagrams. Potential-pH diagrams for the iron/sulfur/water

20 GEOTHERMAL SCALING AND CORROSION

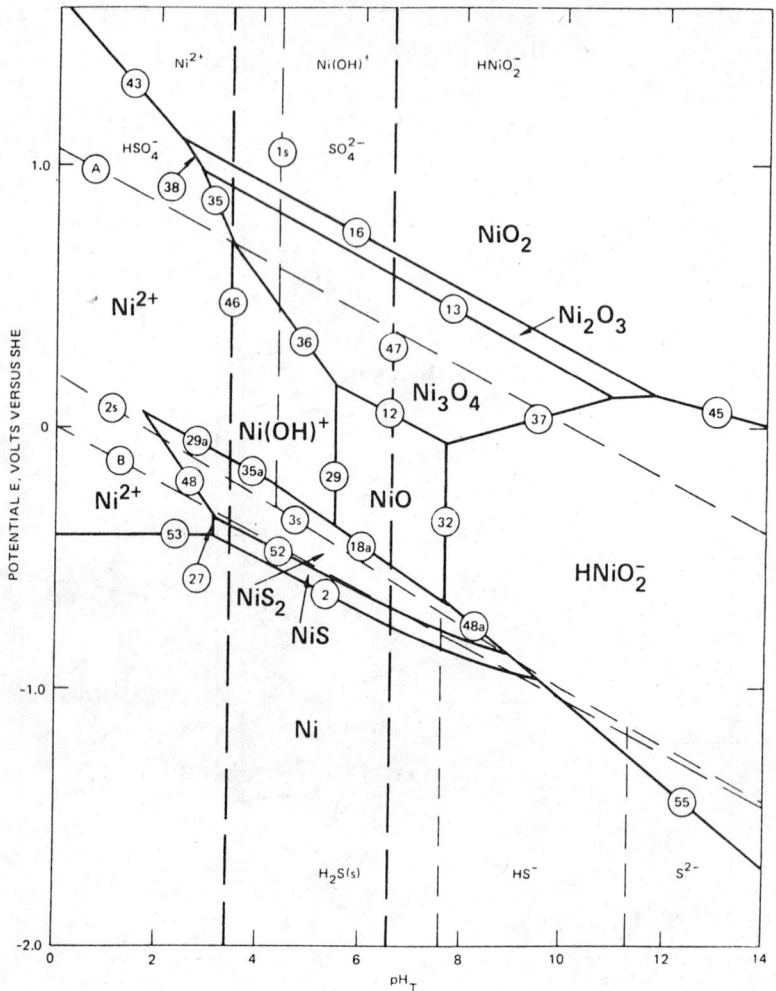

FIG. 4—*Potential-pH diagram for nickel in high-salinity brine at 250°C in the presence of 10 ppm total dissolved sulfide (H_2S + HS^- + S^{2-}). Activities of HSO_4^-, SO_4^{2-}, and dissolved nickel species = 10^{-6} molal. (S) indicates a soluble molecular species.*

system at elevated temperatures in the absence of chloride ion have been reported elsewhere [13,18,19], and, where comparison is possible, the diagrams reported here are in good agreement with those previously published.

In the cases of both iron and nickel at 25°C, the most stable solid oxidation phase is the M(II) sulfide. Only troilite (stoichiometric FeS) was considered in the derivation of the diagram for iron, although it is recognized that other sulfide phases, such as mackinawite ($Fe_{1+x}S$) and pyrrhotite

(FeS_{1+x}) exist, and indeed mackinawite may form at lower potentials than troilite under certain conditions [13]. As the potential is increased, oxidation of MS to MS_2 is predicted to occur over almost the entire pH range. Note that iron disulfide (FeS_2) and nickel disulfide (NiS_2) are Fe(II) and Ni(II) phases, respectively; the change in oxidation state is associated with the sulfur anions in the lattices. Thus, FeS_2 and NiS_2 are best described as iron and nickel disulfides in which the average oxidation state of sulfur is −1. Note also that the conversion of MS to MS_2 involves reaction with H_2S (or its anions HS^- and S^{2-}), as shown by the location of the equilibrium Lines 11 (Fig. 1) and 52 (Figs. 3 and 4) in the stability region for sulfide (H_2S and HS^-) in the potential-pH diagram for the sulfur/water system. On the other hand, equilibrium between MS_2 and the oxides Fe_3O_4, Fe_2O_3, and NiO involves sulfur oxyanions as demonstrated by the potentials for these processes lying within the stability regions for HSO_4^- and SO_4^{2-}.

Increasing the temperature from 25 to 250°C proved to have several important consequences for the thermodynamic equilibrium behavior of iron, nickel, and sulfur in geothermal brine. Thus, the stable iron III complex ($FeCl_2^+$) at 25°C is no longer the predominant species at 250°C; the decrease in the dielectric constant of the medium favors the formation of the neutral complex $FeCl_3$. Similarly, pK_a for the dissociation of bisulfate ion (HSO_4^-) increases from approximately 1.98 at 25°C to 4.39 at 250°C. Increasing the temperature also has a marked effect on the stability regions for the dissolved metal-containing ions in geothermal brine. For example, the stability regions for cations at low pH values become more restricted as the temperature increases, whereas those for the anions $HFeO_2^-$ and $HNiO_2^-$ [or, equivalently, $Fe(OH)_3^-$ and $Ni(OH)_3^-$] increase. These predicted changes in the stability domains for dissolved species have been noted previously [2-4,13-17], and in the case of the anions the shifts in the lines for equilibrium between the ions and the oxides (for example, Line 32, Fig. 3) appear to be greater than can be accounted for by the change in the dissociation constant of water alone. This observation suggests that the anions become stabilized in relation to the cations at elevated temperatures.

Probably the most important change in the diagrams due to increasing temperature occurs in the relative stabilities of the sulfides and oxides. Thus, in both cases the disulfides FeS_2 and NiS_2 are predicted to exhibit domains of stability at 25°C that extend to very low pH values in the cation stability regions. The lower boundaries of these regions are determined by equilibrium between MS_2 and dissolved M^{2+} and H_2S, whereas the upper boundaries represent equilibrium between MS_2 and dissolved M^{2+} and the oxyanions HSO_4^- or SO_4^2. Thus, if the potential is first increased from within the lower stability region for M^{2+}, at some point the formation of solid FeS$_2$ (Line 37, Fig. 1) or NiS$_2$ (Line 48, Fig. 3) becomes spontaneous by oxidative deposition. However, if the potential is increased still further, oxidative dissolution of the disulfides can occur. These relationships are of considerable importance in the mining and geothermal industries, because they provide the thermodynamic boundaries for the oxidative and reductive dissolution of pyrite ores, for example, and pyrite-rich scales in geothermal systems. At elevated temperatures, however, the stability domains for the sulfides, and particularly for the disulfides FeS$_2$ and NiS$_2$, are reduced sharply in size, thereby indicating much more restrictive conditions for the formation of these phases in high-temperature geothermal systems. Nevertheless, Seward [14] has argued that the downhole redox potential and pH in the Broadlands geothermal field south of Reporoa, New Zealand, lie within the stability region for FeS$_2$, and, indeed, pyrite is observed in core samples from the steam-producing formation.

Acknowledgments

Partial financial support for this work provided by the U.S. Bureau of Mines, under Contract No. J0188076, is gratefully acknowledged.

References

[1] Macdonald, D. D., Syrett, B. C., and Wing, S. S., *Corrosion*, Vol. 35, 1979, p. 1.
[2] Macdonald, D. D., Shierman, G. R., and Butler, P., "Thermodynamics of the Water and Copper-Water Systems," Report No. AECL-4136, Atomic Energy of Canada Ltd., Pinawa, Manitoba, Canada, 1972.
[3] Macdonald, D. D. and Butler, P., *Corrosion Science*, Vol. 13, 1973, p. 259.
[4] Pound, B. G., Macdonald, D. D., and Tomlinson, J. W., *Electrochimica Acta*, Vol. 24, 1979, p. 929.
[5] Naumov, G. B., Ryzhenko, B. N., and Khodakosky, I. L., *Handbook of Thermodynamic Data*, USGS Transl. USGS-WRD-74-001, U.S. Geological Survey, 1974.
[6] *National Bureau of Standards, Technical Note*, No. 270-1, 1965.
[7] *National Bureau of Standards, Technical Note*, No. 270-4, 1969.
[8] Kelley, K. K., *Bulletin of the United States Bureau of Mines*, Vol. 584, 1960.
[9] Wicks, C. E. and Bock, F. E., *Bulletin of the United States Bureau of Mines*, Vol. 605, 1963.
[10] Criss, C. M. and Cobble, J. W., *Journal of the American Chemical Society*, Vol. 86, 1964, pp. 5385, 5390.
[11] Murray, R., and Cobble, J. W., personal communication, 1977.
[12] Taylor, D. F., *Journal of the Electrochemical Society*, Vol. 125, 1978, p. 808.
[13] Macdonald, D. D. and Hyne, J. B., "The Thermodynamics of the Iron/Sulfur/Water System," Final report to Atomic Energy of Canada, Ltd., Pinawa, Manitoba, Canada, 1976.
[14] Macdonald, D. D., Shierman, G., and Butler, P., "The Thermodynamics of Metal-Water Systems at Elevated Temperatures: II. The Iron-Water System," Report No. AECL-4137, Atomic Energy of Canada, Ltd., Pinawa, Manitoba, Canada, 1972.
[15] Macdonald, D. D., "The Thermodynamics of Metal-Water Systems at Elevated Temperatures. IV. The Nickel-Water System," Report No. AECL-4139, Atomic Energy of Canada, Ltd., Pinawa, Manitoba, Canada, 1972.

[16] Syrett, B. C., Macdonald, D. D., Shih, H., and Wing, S. S., "Corrosion Chemistry of Geothermal Brines: Part 1: Low-Salinity Brine," and "Part 2: High-Salinity Brine," Final Report to National Science Foundation (RANN) and Department of Energy, Washington, D.C., NSF(RANN) Grant No. AER 76-00713, 1977.
[17] Macdonald, D. D. in *Modern Aspects of Electrochemistry*, Vol. 11, J. O'M. Bockris and B. E. Conway, Eds., Plenum Press, New York, 1975, p. 141.
[18] Biernat, R. J. and Robins, R. G., *Electrochimica Acta*, Vol. 17, 1972, p. 1261.
[19] Seward T. M., *American Journal of Science*, Vol. 274, 1974, p. 190.

Marshall Conover,[1] *Peter Ellis,*[2] *and Anne Curzon*[3]

Material Selection Guidelines for Geothermal Power Systems—An Overview

REFERENCE: Conover, Marshall, Ellis, Peter, and Curzon, Anne, "**Material Selection Guidelines for Geothermal Power Systems—An Overview,**" *Geothermal Scaling and Corrosion, ASTM STP 717,* L. A. Casper and T. R. Pinchback, Eds., American Society for Testing and Materials, 1980, pp. 24-40.

ABSTRACT: Perhaps the most important difference between traditional electric power generation and geothermal power generation is the potentially severe corrosion of metals caused by the use of the geothermal fluids. The object of this overview is to present the principal results of work conducted for the U.S. Department of Energy under Contract EG-77-C-04-3904. Process streams are identified by the presentation of nine geothermal power cycles applicable to four types of liquid-dominated geothermal resources found in the United States. Of the many constituents in geothermal fluids, seven key chemical species are identified that account for most corrosion phenomena in geothermal power systems. These species are: oxygen, hydrogen sulfide, carbon dioxide, ammonia, chloride, sulfate, and hydrogen ion concentrations. Based on analyses of actual geothermal materials test data and test methods, the performance of metals in geothermal fluid, steam, and condensate is presented for carbon and stainless steels, titanium, nickel, copper, and many other alloys. The applicability of new nonmetallic materials in geothermal systems is also addressed. Finally, the similarities and differences between seawater and geothermal corrosion phenomena are discussed.

KEY WORDS: geothermal, scaling, corrosion, materials, metals, nonmetals, key chemical species, guidelines, process streams, seawater

Important to the development of geothermal energy is the effect that corrosion and materials problems can have on production efficiency and downtime. Proper materials selection requires a knowledge of process stream characteristics that can contribute to equipment failure. The chemical composition, temperature, and velocity of a geothermal stream

---

[1] Geothermal programs manager, Radian Corp., Austin, Tex. 78766.
[2] Geothermal corrosion engineer, Radian Corp., Austin, Tex. 78766.
[3] Mechanical engineer, Radian Corp., Austin, Tex. 78766.

vary depending on geothermal source, the power cycle chosen, and even on the point within any given cycle. Results from materials testing, using actual geothermal fluids, as well as operational experience from existing geothermal power plants, can help to provide a basis for materials selection.

Geothermal Power Cycles

A thermodynamic power cycle is the process employed in extracting and utilizing geothermal energy to produce electricity. Methods used in the power cycle to produce steam or other kinds of vapor to drive a turbine depend on characteristics of the geothermal fluid. The four general types of geothermal resources found in the United States are steam-dominated, liquid-dominated, hot dry rocks, and geopressured resources. National interest in electric generation is focused on the first two types of resources, and it is the typical power cycles for these resources which are described here.

Nine generalized power cycles are potentially applicable to steam-dominated and liquid-dominated geothermal resources. They are listed and classified in Table 1 according to whether the source is steam or liquid dominated and whether it is recovered by natural pressure or by pumping.

The power cycles include three major process steps: recovering geothermal fluid from the well, producing steam or other kinds of vapor to drive a turbine, and recovering condensate and noncondensable gases. Power cycles differ mainly in the methods used to generate steam or another vapor to drive the turbine. The *direct expansion cycle* uses unflashed steam from steam-dominated sources. *Single-flash and dual-flash cycles* employ a change in pressure to separate steam from a liquid-dominated source. *Binary cycles* use a second liquid as an intermediate heat transfer medium. In *direct binary cycles*, vapor to drive the turbine is produced from the second liquid by heat exchange with the geothermal fluid. In

TABLE 1—*Potential geothermal power cycles.*

Steam-Dominated Sources
   Direct expansion cycle
Liquid-Dominated Natural Pressure Sources
   Single-flash steam cycle
   Dual-flash steam cycle
   Direct binary cycle
   Flashed steam binary cycle
   Two-phase expander cycle
Liquid-Dominated Pumped Sources
   Direct binary cycle
   Flashed steam binary cycle
   Two-phase expander cycle

flashed steam binary cycles, steam is separated from the geothermal fluid and used to vaporize the second liquid. *Two-phase expander cycles* use a mixture of both vapor and liquid from the geothermal fluid to drive the turbine.

Power cycles are also classified according to the method of recovering geothermal fluid from the well. Liquid-dominated resources can be recovered either by natural pressure or by a downhole pump. Recovery by natural pressure has the advantage that little hardware is required down the borehole, which reduces the cost and complexity of equipment installation and maintenance. Downhole pumping can be used to increase well production, keep noncondensable gases dissolved, and prevent uncontrolled chemical changes and scaling in the wellhead fluid caused by flashing during the recovery process. The economic advantage of higher well flow rates from pumping may offset the cost of installation and maintenance of downhole pumps.

In selecting a power cycle for a particular geothermal site, consideration must be given to scaling potential and the concentration of noncondensable gases in the process stream. The scaling tendencies of a given source depend on its temperature and the concentration of certain chemical species, namely, silica, calcium, carbonate, sulfate, and heavy metal ions. Scaling affects the choice of cycle but requires a number of trade-offs, and its effects cannot be generalized.

Because the energy obtained from noncondensable gases in a turbine is small compared with the energy available from the steam, a turbine that can handle both must be larger and, therefore, more expensive than one with an equivalent rating that handles steam alone. Since noncondensable gases must be continuously removed from the condenser, as the quantity of these gases increases, the pump work required to remove them increases. It is probable that the gas-extracting work may equal the work extracted from the condensing turbine at a gas concentration of 3.5 to 4.0 weight percent. Thus, flashed steam condensing cycles are generally and economically applicable to resources with noncondensable gas concentrations of less than 3 weight percent. Intermediate binary cycles are alternatives to the use of gas ejectors and compressors for resources with higher noncondensable gas concentrations.

Corrosive Species in Geothermal Fluids

Geothermal fluids may contain seven key chemical species that produce a significant corrosive effect on metallic construction materials. Some generalizations about the corrosive effects of these species are discussed here:

Hydrogen Ion—The general corrosion rate of carbon steels increases rapidly with increasing hydrogen ion (decreasing pH), especially below pH 7. The passivity of many alloys is pH dependent. Breakdown of passivity at local areas can lead to serious forms of attack, such as pitting, crevice corrosion, and stress corrosion cracking.

Chloride—Chloride (Cl$^-$) causes local breakdown of passive films, which protect many metals from uniform attack. Local penetration of this film can cause pitting, crevice corrosion, or stress corrosion cracking. Uniform corrosion rates can also increase with increasing chloride concentration, but this action is generally less serious than local forms of attack.

Hydrogen Sulfide—Probably the most severe effect of hydrogen sulfide (H_2S) is its attack on certain copper and nickel alloys. These alloys have performed well in seawater but are practically unusable in geothermal fluids containing H_2S. The effect of H_2S on iron-based materials is less predictable. Accelerated attack occurs in some cases and inhibition in others. High-strength steels are often subject to sulfide stress cracking. Hydrogen sulfide may also cause hydrogen blistering of steels. Oxidation of H_2S in aerated geothermal process streams increases the acidity of the stream.

Carbon Dioxide—In the acidic region, carbon dioxide (CO_2) can accelerate the uniform corrosion of carbon steels. The pH of geothermal fluids and process streams is largely controlled by CO_2. Carbonates and bicarbonates can display mild inhibitive effects.

Ammonia—Ammonia (NH_3) can cause stress corrosion cracking of some copper alloys. It may also accelerate the uniform corrosion of mild steels.

Sulfate—Sulfate ($SO_4^=$) plays a minor role in most geothermal fluids. In some low-chloride streams, sulfate will be the main aggressive anion. Even in this case, it rarely causes the same severe localized attack as chloride.

Oxygen—The addition of part-per-billion quantities of oxygen (O_2) to a high-temperature geothermal system can greatly increase the chance of severe localized corrosion of normally resistant metals. The corrosion of carbon steels is sensitive to trace amounts of oxygen.

Transition Metal Ions—"Heavy" or transition metal ions might also be included as key species. Their action at low concentrations on most construction materials is ill defined. However, the poor performance of aluminum alloys in geothermal fluids may be due in part to low levels of copper or mercury in these fluids. Geothermal fluids from the Salton Sea, Calif., known geothermal resource area (KGRA) contain many transition metal ions at greater than "trace" concentrations. Some oxidized forms of transition metal ions (Fe^{3+}, Cu^{2+}, and others) are corrosive, but these ions are present in the lowest oxidation state (most-reduced form) in geothermal fluids. Oxygen can convert Fe^{2+} to Fe^{3+}, which is another reason to exclude oxygen from geothermal streams.

Typical concentrations of the species just described in fluid from KGRAs are listed in Table 2 [*1*].[4]

---

[4] The italic numbers in brackets refer to the list of references appended to this paper.

TABLE 2—*Typical concentrations of key corrosive chemical species in fluid from seven KGRAs.*

KGRA	Temperature °C and Location	Properties of the Geothermal Resource								Fluid Description[a]	Comments
		pH	Cl⁻	Total CO₂	Total H₂S	Total NH₃	SO₄			

KGRA	Temperature °C and Location	pH	Cl⁻	Total CO₂	Total H₂S	Total NH₃	SO₄	Fluid Description[a]	Comments
Salton Sea, Calif.	250 (borehole)	5.2	115 000	1 000	10 to 30	300	20	unflashed wellhead fluid	these data are the results of extensive analyses of several wells, so they are representative of the resource. Variations occur; CO₂ concentrations as high as 10 000 ppm have been measured.
East Mesa, Calif.	180 to 200 (borehole)	5.7	11 000	800	3	41	20	unflashed wellhead fluid	the data are for Well 6-1, which has received the most study. It may not be typical of the field. Higher pH (6.5) and lower Cl and CO₂ have been measured at other wells. The fluid also contains about 9 ppm Fe.
Heber, Calif.	180 to 200	7.1	9 000	180	~2	13	152	unflashed wellhead fluid	these data are from two separate analyses of fluid from the Well Nowlin 1. The fluid also contains about 4 ppm Fe.
Mono-Long Valley, Calif.	175 (borehole)	6.5	227	180	14	0.1	96	unflashed wellhead fluid	these data are for well Endogenous 4. It is the only well for which H₂S concentrations were measured. Results of analyses of other species in fluids from other wells in the area agree with the results for this well.
Baca (Valles Caldera) N. Mex.	171 (wellhead at 110 psig)	6.8	3 770	128	6	...	59	flashed fluid	these data represent the results of three similar analyses of flashed fluid from well Baca 11. They are not representative of values in the unflashed fluid.
Beowawe, Nev.	132 (wellhead)	9.3	50	209	6	3	89	flashed fluid	these data are representative of wells Vulcan 2, 3, and 4. Data for H₂S, CO₂, NH₃ and pH are not representative of the unflashed fluid.
Raft River, Ind.	146 (borehole)	7.2	780	60	0.1	2	61	unflashed fluid	these data are for Well RRGE 1 which is well characterized. Well RRGE 2 has a similar composition. Well RRGE 3 has a similar composition except for Cl⁻, which is about three times higher.

[a] Because these measurements were made at different points (i.e., before or after flashing), the source fluids cannot be directly compared. Often during flashing, chloride concentration increases while CO₂ and H₂S decrease. An increase in pH will generally result.

Scaling or solids deposition is another aspect of geothermal fluid chemistry that influences materials performance. Precipitation of liquid phase species in solution or on equipment surfaces can influence corrosion rates and cause erosion. The composition of the scale-forming solids and the rate of precipitation depend on the fluid composition and specific process stream conditions; therefore, scale-forming species are not included in the list of key species.

Weighing the relative significance of these species is difficult for several reasons. First, the aggressiveness of a particular species varies from one material to the next. Often, the interaction of two or more species on materials has a different effect from that of each species alone. Also, the temperature dependence of corrosion of a given material by a given species is often undefined. Finally, the importance of a given species depends on the form of attack under consideration.

Corrosion Modes for Metals in Geothermal Systems

Corrosive attack of metals can occur in several of the following forms [2]:

Uniform Corrosion—This is a general, allover attack of the metal surface. Uniform corrosion is often promoted by chloride, carbon dioxide, oxygen, or ammonia.

Pitting—Pitting is a localized form of attack, which results in the development of small pits in the metal surface. Pitting is often associated with the breakdown of a protective film or surface scale.

Crevice Corrosion—Crevice corrosion is similar to pitting in that it is a localized attack. Unlike most other forms of corrosion, it is geometry dependent and forms in the crevices of equipment.

Stress Corrosion Cracking—Stress corrosion cracking (SCC) is a catastrophic type of failure promoted by a combination of tensile stress and the presence of chloride ion in the environment. The presence of oxygen and increasing temperature increase the severity of attack.

Sulfide Stress Cracking—Sulfide stress cracking (SSC) is a catastrophic failure that results from exposure under stress of susceptible materials to environments containing H_2S in an aqueous phase. In contrast to stress corrosion cracking, SSC decreases in severity with increasing temperature, but oxygen may have little, if any, effect on the SSC mechanism. On the other hand, low pH greatly accelerates failure.

Hydrogen Blistering—Rupture of metallic materials results when hydrogen trapped in voids accumulates at a sufficient pressure. The material need not be under stress for hydrogen blistering to occur.

Intergranular Corrosion—Intergranular corrosion is preferential corrosion at or adjacent to grain boundaries, with little or no attack on the bodies of the grains. The alloy disintegrates (grains fall out) or loses its strength, or both.

Galvanic Corrosion—This occurs when two metals are electrically connected. Corrosion of the less noble material will be accelerated. Materials may be ordered in a galvanic series (by increasing nobility) to help in materials selection. Care must be taken because the order of metals may change with variations in chemistry and temperature.

Corrosion Fatigue—Corrosion fatigue is premature fracture when cyclic stresses are imposed on a material in a corrosive environment. The corrosion fatigue limit is the greatest unit stress that may be applied under given conditions of stress, rate of stress application, temperature, and corrosive environment without causing the material to fail in a given number of cycles of stress [3]. The combined effects of cyclic stress and corrosion are often far more severe than the simple sum of their actions.

Exfoliation—This involves the formation of discrete layers of corrosion products or of metal separated from the lattice by corrosion products. The layers may break loose, damaging downstream components.

Performance of Metals in Geothermal Systems

Operating experience and field testing in a variety of geothermal applications can lead to generalizations of material performance. These generalizations are discussed in the following three sections for metals exposed to liquid, condensate, and steam geothermal streams.

Performance of Metals in Wellhead and Flashed Liquid Streams

In addition to geothermal operating experience and field tests with geothermal fluids, materials performance data from seawater distillation plants has been considered. (Seawater is somewhat comparable to some geothermal fluids as far as dissolved solids are concerned.) The primary corrosion modes for each of several classes of metals are discussed here and summarized in Table 3 [4].

Mild and Low-Alloy Steels

The low cost, availability, and ease of fabrication of low-carbon steels (mild steels) make them attractive construction materials for geothermal power plants. However, the reliability of these steels depends upon their applications in the power plants. By taking appropriate precautions, mild steels can be used for thick-walled applications in contact with most geothermal fluids. Thin-walled applications will be limited by the susceptibility of these materials to localized attack, such as pitting and crevice corrosion.

Uniform and Localized Corrosion—Uniform and localized corrosion are the two main modes of corrosion of low-strength mild steels in geothermal systems. Results of a variety of geothermal field tests indicate that uniform corrosion rates are generally about 0.03 to 0.3 mm per year when the pH is greater than 6 and the chloride concentration is less than 2 percent. (Rapid increase of corrosion rate occurs below pH 6 and above 2 percent chloride.) Localized corrosion occurs to some extent in most fluids and becomes predominant in fluids where uniform corrosion is less severe. The chloride ion is the main initiator of localized attack. Hydrogen sulfide can increase the severity of localized corrosion.

Several operating conditions can accelerate the corrosion of mild steels. The introduction of parts-per-billion quantities of oxygen can greatly increase uniform corrosion and initiate pitting and crevice corrosion. High flow rates, in conjunction with entrained solids in the stream, can cause erosion-assisted corrosion. The best flow rate for the use of carbon steels is in the 1.5 to 2 m/s range.

The scales formed on steel by precipitation from geothermal fluids are porous and prone to cracking. Corrosive attack can occur at these small exposed areas, particularly if the steel is galvanically coupled to a more noble metal.

Mill scale left on steel can accelerate localized corrosion, especially in the presence of chlorides. Protective coatings should be used to minimize uniform and localized attack of exterior surfaces.

Sulfide Stress Cracking—Sulfide stress cracking can result in brittle failure of high-strength alloys exposed to aqueous H_2S while under stress. Resistance to SSC generally increases with increasing temperature, decreasing stress, decreasing yield strength, decreasing H_2S concentration, and increasing pH.

Hydrogen Blistering—Hydrogen blistering occurs in low-strength steels exposed to aqueous solutions containing H_2S and has been a problem at the Wairakei, New Zealand, geothermal plant [5]. Since voids are required for blistering, void-free (killed) steels can resist blistering, although manufacturing processes can introduce other blister sites. Though not necessarily subject to blistering, voids in welds may accumulate molecular hydrogen and burst. Stress is not required for hydrogen blistering.

Stainless Steels

The uniform corrosion rate of most stainless steels in geothermal fluids is low, but many are subject to the more serious forms of corrosion: pitting, crevice corrosion, stress corrosion cracking, sulfide stress cracking, intergranular corrosion, and corrosion fatigue. Stainless steels have been used in geothermal environments, but care must be taken in their selection and application.

Pitting and Crevice Corrosion—Crevice corrosion can be a serious problem in stainless steels because they are frequently used in complex

TABLE 3—*Forms and causes for metals in liquid geothermal streams and ways to prevent attack.*

Material	Major Forms of Corrosion	Main Environmental Factors	Limits and Precautions	Other Comments
Mild and low alloy steels	uniform	pH chloride flow velocity	rapid rate increase below pH 6 rapid rate increase above 2 percent Cl^- limit flow to 5 to 7 fps	air in-leakage is a major hazard; local flashing in pipes can cause very high flowrates and erosion/corrosion; avoid direct impingement on steel
	pitting, crevice	temperature chloride scale	susceptibility increases with increasing temperature and chloride concentration remove mill scale; avoid deposits	avoid mechanical crevices
	sulfide stress cracking	H_2S yield strength (hardness) temperature	can occur at very low H_2S levels use low strength material wherever possible hazard greater at lower temperatures	complex interactions
	hydrogen blistering	H_2S	use void-free materials	possible at very low H_2S concentrations more severe when material has porous coating or scale
	galvanic coupling	electrical contact with more noble metal	avoid coupling close to large area of cathodic metal	
Stainless steels Ferritic alloys	pitting, crevice	chloride	in general susceptibility increases with increasing concentration and temperature	lower alloys may also have high uniform rates in severe environments; O_2 is a hazard; higher alloys are much more resistant; Cr and Mo most effective alloying agents
		scale stagnant or low flow oxygen	avoid scale deposits avoid stagnant or low flow conditions O_2 greatly increases susceptibility	
	intergranular	chloride, temperature	avoid by proper welding and heat treating procedures	
Austenitic alloys	stress corrosion cracking	chloride oxygen temperature	complex interaction; depending on other factors, cracking can occur for $Cl^- > 5$ ppm; $O_2 \sim 100$ ppb; $T > 60°C$	hazard increases with increase in Cl^-, O_2, T; some alloys more resistant; protect exterior surfaces
	pitting, crevice	chloride, temperature	see ferritics above	resistance increase with Mo content; avoid mechanical crevices

Martensitic alloys	intergranular	scale stagnant or low flow oxygen chloride, temperature	avoid scale deposits avoid stagnation or low flow conditions O_2 greatly increases susceptibility avoid by proper welding and heat treating procedures
Cast alloys	as above sulfide stress cracking as above	as above H_2S, temperature, stress, hardness	as above more severe at lower temperatures; use low strength levels where possible	general corrosion resistance depends on composition
Titanium alloys	crevice, pitting	chloride, temperature, pH	maximum temperature for resistance depends on chloride and pH	see comments for equivalent wrought alloy; good crevice corrosion resistance needed for pumps and valves
	galvanic coupling	electrical contact with more active metal	coupling to large area of more active metal may cause hydrogen embrittlement of Ti	several alloys have much better resistance than pure Ti; precracked Ti may undergo stress corrosion cracking
Nickel alloys	crevice, pitting	chloride, temperature	similar to stainless steels except higher alloys more resistant to crevice corrosion; high flow rates	
Copper alloys	pitting, uniform, dealloying stress corrosion cracking	H_2S, chloride, temperature ammonia, pH	H_2S as low as 0.1 ppm can cause attack pH and alloy dependence	resistance depends on alloy composition; may be susceptible to hydrogen embrittlement when coupled to steel
Other metals Cobalt alloys			avoid galvanic coupling to steel or other active metal	usefulness limited in H_2S environment
Zirconium and tantalum				several alloys have good sulfide stress cracking resistance at high strength
Aluminum	pitting, crevice	Hg and Cu ions, pH, chloride, temperature	poor results obtained in geothermal tests	resistant to low pH, hot chloride solutions may be useful as exterior siding and construction material

equipment. Susceptibility to localized corrosion increases as the chloride content of the environment increases. The electrochemical pitting potential decreases with increasing temperature, which increases the risk of pitting.

The pitting and crevice corrosion resistance of stainless steels is strongly dependent on their chromium and molybdenum contents, as tests in deaerated artificial seawater indicate [6].

Stress Corrosion Cracking—Austenitic stainless steels are susceptible to SCC in hot chloride solutions. Ferritic stainless steels are generally more resistant. Stress corrosion cracking depends on the chloride and oxygen concentrations, pH, temperature, stress, and alloy composition.

The severity of cracking increases with temperature. For any given concentration of chloride, oxygen, and pH, a "lower critical temperature" exists below which SCC does not occur. Stress is required for SCC, but there is no evident lower critical applied stress. The time to fracture decreases sharply at total stress equal to or above the yield point. The total stress is the critical factor. Applied stress actually presents relatively little design difficulty because it can be quantified and allowed for. Residual stresses from cold work and welding are unpredictable. Postwelding stress may be near the yield point. Failures have occurred at zero applied stress [7]. Austenitic stainless steels should be fully annealed, and subsequent applied stresses should be minimized.

Nickel is the constituent that has the major effect on susceptibility of stainless steels to SCC in chloride solutions. Immunity from chloride stress cracking is not normally obtained unless the nickel content is less than 1 percent or greater than 45 percent [7,8]. Additions of molybdenum and silicon improve resistance to SCC [9,10,11].

Intergranular Corrosion—Intergranular corrosion occurs in austenitic stainless steels when they are sensitized (that is, heated in the 510 to 790°C range) [12]. When stainless steels are welded, a zone called the heat-affected zone (HAZ) along each side of the weld is heated into the sensitizing range. The HAZ is a site of intergranular attack, often called weld decay.

Intergranular corrosion can be controlled in three ways. First, the material can be quench-annealed or solution-annealed (heated to high temperature, typically 1065 to 1120°C, and water quenched). HAZ intergranular corrosion can be controlled by lowering the carbon content to below 0.03 percent. The addition of stabilizers such as niobium or titanium can also help prevent intergranular attack [13].

Ferritic stainless steels are also subject to intergranular corrosion, although the range of sensitizing temperatures has not been established. The preventive measures just described for austenitic steels generally apply to ferritic stainless steels as well.

Sulfide Stress Cracking—Martensitic and ferritic stainless steels are very susceptible to SSC, whereas austenitic steels tend to be immune. Sulfide stress cracking is more severe at lower temperatures. Low-strength steels are more resistant to SSC.

Titanium and Titanium Alloys

Titanium and its alloys have given good results in all but the most extreme environments when tested for geothermal applications. Titanium was used successfully for hydrogen and oil coolers exposed to aerated cooling water and condensate at the geothermal facility in Cerro Prieto, Mexico [14]. Two other heat-exchanger materials had failed in this environment.

Uniform Corrosion—The uniform corrosion rate of titanium and titanium alloys tested in geothermal fluids has been less than 0.008 mm per year, and under even the worst conditions has not exceeded 0.13 mm per year.

Increasing temperature and chloride concentration do not increase uniform corrosion of titanium alloys. Experience in seawater desalination plants indicates that fluid velocities of 9 m/s have no effect on uniform corrosion. Titanium is also very resistant to impingement and cavitation damage [15].

Pitting and Crevice Corrosion—No significant local corrosion should occur in environments with less than 10 percent chloride. Experience in desalination and the chemical process industry shows that corrosion in tight crevices may be a problem in high-temperature, high-chloride solutions similar to Salton Sea fluid. Pitting may be an occasional problem.

Some titanium alloys are much more resistant to local corrosion than commercially pure titanium. The alloys Ti-0.2Pd and TiCode-12 show particularly good resistance.

Galvanic Coupling and Hydrogen Embrittlement—Titanium is cathodic to most other metals in saline environments. If the titanium area is large compared with the metal to which it is coupled, the second metal may corrode severely. Since titanium is the cathode in most galvanic couples, hydrogen can be formed on titanium coupled to an active metal. Titanium is capable of absorbing hydrogen and becoming embrittled [16]. Some titanium-nickel alloys are susceptible to hydrogen embrittlement in desalination service [17].

Stress Corrosion Cracking—Commercial titanium alloys are not considered to be susceptible to cracking in 3 percent sodium chloride (NaCl) solutions at ambient temperatures. However, precracked specimens of many such alloys fail rapidly in 3 percent NaCl solutions. This behavior indicates that the resistance of some titanium alloys is dependent on the integrity of the protective oxide film and not on the intrinsic resistance of the alloy lattice [18]. Stress corrosion cracking should, therefore, not be a major problem, provided that the components are not precracked during fabrication and that the design does not allow excessive vibration of the titanium parts.

The susceptibility of precracked titanium alloys to SCC is adversely affected by the aluminum, tin, manganese, cobalt, and oxygen content of the alloy. Some alloys containing more than 6 percent aluminum are especially susceptible. Additions of molybdenum, niobium, or vanadium reduce or eliminate the susceptibility of titanium alloys to chloride-enhanced propagation of cracks in precracked specimens [19].

Nickel-Based Alloys

High-nickel alloys are frequently used to combat severe corrosion problems. The nickel-chromium-molybdenum alloys appear to be the most applicable to high-temperature geothermal fluids. Inconel 625 and Hastelloy C-276 are particularly resistant to corrosion. These alloys can tolerate very high flow rates and occasional aeration. Similar alloys containing iron in place of molybdenum face competition from the most resistant stainless steels, but may find application where their mechanical properties are desirable. Cupronickels will have limited usefulness in geothermal streams containing even trace quantities of H_2S.

Copper-Based Alloys

The use of copper alloys in geothermal fluids is severely limited by the relatively high concentrations of sulfide found in most sources. The Raft River, Idaho, KGRA, with a low sulfide concentration of 0.1 ppm, appears to be an exceptional case. Even in this fluid, the performance of copper-nickel alloys has been very poor. Although some nickel-free brasses and bronzes have performed well in testing, dealloying of some copper alloys has been observed [20].

The cracking of some copper-based alloys while exposed to ammonia or ammonia derivatives is known as "season cracking" [21] or ammonia SCC. Provided NH_3 and NH_4^+ levels are low, cracking by ammonia species in the liquid phase of geothermal fluid should be limited to stagnant areas. Susceptibility of copper-zinc brasses to cracking increases with increasing zinc concentration.

Other Metallic Materials

A number of other metals and alloys either may have important specialized but limited uses in geothermal applications or have not shown promise in geothermal tests.

Cobalt Alloys—These alloys may find application in forms of service requiring high strength combined with resistance to sulfide stress cracking and in forms of service requiring wear resistance.

Zirconium and Tantalum—These metals may be considered for service in severe, hot, acid chloride environments, such as in injection nozzles for acidifying fluid with hydrochloric acid.

Aluminum Alloys—These alloys have not shown good resistance in tests conducted in direct contact with geothermal fluids. Pitting and attack caused by galvanic coupling are particularly severe.

Performance of Metals in Condensate Streams

Although much of the information given previously for liquid streams is also applicable to condensate and steam environments, there are some additional considerations for these streams. The corrosivity of condensate streams depends to a large extent on the following interacting factors:

1. The efficiency of steam separation and thus the chloride content of the condensate.
2. The quantity of the flashed noncondensable gases, CO_2, H_2S, and NH_3, that are absorbed by the condensate.
3. The pH of the condensate resulting from absorption of these gases.
4. Aeration or mixing with aerated water.

The effects on metals of pH, H_2S, CO_2, NH_3, and oxygen, described previously for liquid streams, are applicable here. Although condensate streams have relatively low chloride concentrations, their potential corrosivity should not be underestimated. For example, assuming 25 percent flashing of a wellhead liquid containing 1 percent chloride and 0.05 percent liquid carry-over, the condensate would contain about 20 ppm chloride. This is approximately equal to the minimum chloride concentration needed to cause pitting of plain steels at 25°C [22]. Aeration may cause corrosion rates in steel of several millimetres per year. If the condensate is aerated, and the temperature is greater than 60 to 80°C, stress corrosion cracking of austenitic stainless steels could occur at this chloride concentration.

Performance of Metals in Contact with Steam from Liquid-Dominated Sources

Steam from liquid-dominated geothermal fluids has corrosion characteristics of its own, but its general corrosivity is dominated by two properties common to the saline fluid and condensate streams:

1. Carry-over of entrained liquid provides the chloride needed for localized attack. High-velocity impingement of droplets is conducive to localized attack. Efficient steam separation and steam scrubbing are

important, but may not preclude attack. For a given steam separation and scrubbing efficiency, corrosion may depend on the chloride content and corrosivity of the liquid stream.

2. Areas where local condensation may occur are subject to attack by low-pH condensate containing H_2S, CO_2, and some chloride. The most noteworthy of these locations are the low-pressure turbine section, liquid traps, and poorly insulated or stagnant parts of steam transfer sections. These last locations should be eliminated by plant design.

Corrosion Fatigue—Corrosion fatigue is a potential problem in turbines driven by geothermal steam. Tests on carbon steels and low-alloy steels in fresh water, salt water, and seawater show that fatigue endurance limits in a moderately corrosive medium (for example, salt water) are practically independent of the chemical composition of the steel [23]; however, variations in the steam composition can have a significant impact on the endurance limit. Increased concentrations of O_2, CO_2, bicarbonate (HCO_3^-), and H_2S severely reduce the endurance limit, while often causing sulfide stress cracking, pitting, or hydrogen blistering as well. Chloride appears to have no significant effect on the fatigue endurance of low-alloy steels.

Stainless steels are more resistant to corrosion fatigue than the low-alloy and carbon steels. Chromium is the most effective alloying element in the absence of H_2S, but nickel is more effective against H_2S [24]. The best results are obtained with a combination of chromium and nickel [25]. Molybdenum is also beneficial as an alloying element [26].

Exfoliation—Flakes of iron sulfide scale from steam lines are another potential source of damage to turbines operating directly on flashed steam. Iron sulfide coatings form on steel pipes carrying steam containing H_2S. There is a tendency for these coatings to crack and flake off. If this occurs in lines upstream of the turbine, the scale particles can be carried into the turbine, causing erosion and possibly erosion–corrosion damage.

A much slower flaking of magnetite scales has caused serious damage to turbines in high-temperature, high-pressure fossil fuel plants. This experience accentuates the need for close monitoring and frequent inspection of geothermal turbines.

Nonmetallic Materials

The search for construction materials in geothermal environments has concentrated primarily on metallic materials, but applications of nonmetallic materials are receiving increasing attention. Nonmetallic materials are required in some geothermal operations, such as elastomers in drilling operations. In other areas, these materials may be cost-effective replacements for metallic materials. Nonmetallic materials have some advantages. They are generally resistant to corrosion at conditions that may adversely affect metals and alloys, and their installation costs may be lower than those for metals. However, nonmetallic materials are subject to degradation, and geothermal fluids severely test their durability. They are not useful in heat-transfer equipment.

The performance of nonmetallic materials in geothermal environments is in the early stages of investigation, and the test results are somewhat limited. Some comparisons and trends may be found by examining the results of using nonmetallic materials in desalination fluids and in conventional drilling technologies. Nonmetallic materials that may find geothermal applications are discussed in the following paragraphs.

Concrete Polymer Composites—The durability of polymer concrete is dependent upon the aggregate composition. At temperatures above 218°C, only polymer concrete materials containing mixtures of silica sand and portland cement have been resistant to geothermal brine and steam [27].

Cements—Polymer concretes and inorganic cements such as calcium-silicate cements and phosphate-bonded glass cements are being investigated as potential cementing materials. Corrosion-resistant cements may also find applications as lining materials.

Elastomers—Research is presently being conducted to evaluate the application of elastomers as packer seals and drill-bit cone cutter seals [28].

Fiber-Reinforced Laminates—Fiber-reinforced plastic (FRP) laminates may be useful because of their high corrosion resistance and reasonable cost. Their potential applicability is based on tests in desalination environments, and testing in geothermal environments is required.

Additional materials that are potentially applicable for geothermal use include thermoplastics, fiber-reinforced plastic and coated pipe, and paint and coatings.

Acknowledgment

The basic work for this study was conducted under U.S. Department of Energy (DOE) Contract No. EG-77-C-04-3904. We wish to thank Dr. Robert R. Reeber, program manager in the Division of Geothermal Energy, at the U.S. Department of Energy, for his guidance and assistance.

References

[1] DeBerry, D. W., Ellis, P. F., and Thomas, C. C., *Materials Selection Guidelines for Geothermal Power Sysetm: First Edition*, Document No. ALO/3904-1, Radian Corp., Austin, Tex., Sept. 1978, pp. 3-7.
[2] DeBerry, D. W., Ellis, P. F., and Thomas, C. C., *Materials Selection Guidelines for Geothermal Power System: First Edition*, U.S. Department of Energy, Document No. ALO/3904-1, Radian Corp., Austin, Tex., Sept. 1978, pp. 5-1 ff.
[3] Wescott, B. B., *Mechanical Engineering*, Vol. 60, 1938, p. 813.
[4] DeBerry, D. W., Ellis, P. F., and Thomas, C. C., *Materials Selection Guidelines for*

Geothermal Power System: First Edition, U.S. Department of Energy, Document No. ALO/3904-1, Radian Corp., Austin, Tex., Sept. 1978, p. 5-3.
[5] Marshall, T. and Tombs, A., *Australasian Corrosion Engineering*, Vol. 13, No. 9, 1969, p. 7.
[6] Pessal, N. and Nurminen, J. I., *Corrosion*, Vol. 30, 1974, p. 381.
[7] Latanision, R. M. and Staehle, R. W., "Stress Corrosion Cracking of Iron-Nickel-Chromium Alloys," *Proceedings*, Conference on Fundamental Aspects of Stress Corrosion Cracking, Ohio State University, Columbus, Ohio, Sept. 1967, pp. 214 ff.
[8] Watkins, M. and Green, J. B., *Journal of Petroleum Technology*, June 1976, pp. 698-704.
[9] Lizlovs, E. A., *Journal of the Electrochemical Society*, Vol. 124, No. 12, 1977, p. 1887.
[10] Dundas, H. J., "Effect of Molybdenum on Stress Corrosion Cracking of Austenitic Stainless Steel," Climax Molybdenum Co., Ann Arbor, Mich., Sept. 1975.
[11] Loginow, A. W., Bates, J. F., and Mathay, W. L., *Materials Performance*, Vol. 11, No. 5, 1972, p. 35.
[12] Fontana, G. and Greene, N. D., *Corrosion Engineering*, McGraw-Hill, New York, 1967.
[13] Fontana, G. and Greene, N. D., *Corrosion Engineering*, McGraw-Hill, New York, 1967.
[14] Mañón, Alfredo M., *Proceedings*, Second Workshop on Materials Problems Associated with the Development of Geothermal Energy Systems, El Centro, Calif., May 1975, BuMines Grant No. 1520 88, Geothermal Resources Council, Davis, Calif., 1976, pp. 69-85; available from National Technical Information Service as No. PB 261 349.
[15] George, P. F., Manning, J. A., Jr., and Schrieber, C. F., *Desalination Materials Manual*, U.S. DOE Contract 14-30-3244, U.S. DOE/Dow Chemical Co., Freeport, Tex., May 1975.
[16] Covington, C., *Metal Progress*, Feb. 1977, pp. 38-55.
[17] George, P. F., Manning, J. A., Jr., and Schrieber, C. F., *Desalination Materials Manual*, U.S. DOE Contract 14-30-3244, Dow Chemical Co., Freeport, Tex., May 1975.
[18] Boyd, W. K., "Stress Corrosion Cracking of Titanium and Its Alloys," *Proceedings*, Conference on Fundamental Aspects of Stress Corrosion Cracking, Ohio State University, Columbus, Ohio, Sept. 1967, pp. 593 ff.
[19] Boyd, W. K., "Stress Corrosion Cracking of Titanium and Its Alloys," *Proceedings*, Conference on Fundamental Aspects of Stress Corrosion Cracking, Ohio State University, Columbus, Ohio, Sept. 1967, pp. 593 ff.
[20] Miller, R. L., "Results of Short-Term Corrosion Evaluation Tests at Raft River," TREE-1176, U.S. Department of Energy Contract EY-76-C-07-1570, E G & G Idaho, Inc., Idaho Falls, Idaho, Oct. 1977.
[21] Fontana, G. and Greene, N. D., *Corrosion Engineering*, McGraw-Hill, New York, 1967.
[22] Szklarska-Smialowska, Z., *Corrosion*, Vol. 27, No. 6, 1971, p. 223.
[23] Mehdizadeh, R., McGlasson, L., and Landers, J. E., *Corrosion*, Vol. 22, No. 12, 1966, p. 325.
[24] Wescott, B. B., *Mechanical Engineering*, Vol. 60, 1938, p. 813.
[25] Gilbert, P. T., *Metallurgical Reviews*, Vol. 1, 1956, pp. 379-417.
[26] Wescott, B. B., *Mechanical Engineering*, Vol. 60, 1938, p. 813.
[27] Kukacka, L. E., Fontana, J., Sugama, T., Horn, W., and Amaro, J., "Alternate Materials of Construction for Geothermal Applications," Progress Report No. 13, April-June 1977, BNL 50699, Brookhaven National Laboratory, Upton, N.Y., 1977.
[28] Cassidy, P. E., "Geothermal Elastomer Materials Program Bi-Monthly Meeting Reports," Texas Research Institute, Austin, Tex., 1977-1978.

M. J. Danielson[1]

Application of Linear Polarization Techniques to the Measurement of Corrosion Rates in Simulated Geothermal Brines

REFERENCE: Danielson, M. J., "**Application of Linear Polarization Techniques to the Measurement of Corrosion Rates in Simulated Geothermal Brines,**" *Geothermal Scaling and Corrosion, ASTM STP 717,* L. A. Casper and T. R. Pinchback, Eds., American Society for Testing and Materials, 1980, pp. 41-56.

ABSTRACT: The linear polarization or polarization resistance (PR) technique was investigated in high and low salinity geothermal brine at 150 and 250°C on a low-carbon steel (A53B) and a ferritic stainless steel (E-Brite 26-1) with and without the presence of oxygen in the brine. There was good agreement between the weight loss data and the linear polarization data. Oxygen generally accelerates the corrosion rate, and in the presence of oxygen the usual form of the polarization resistance equation must be replaced by the mass transfer equation to predict the correct corrosion rates. The results were compared with the corrosion rate determined by commercial PR instrumentation, and it was concluded that this instrumentation is adequate for field investigations in which oxygen is absent. However, when oxygen is present, the commercial instruments may underestimate the actual corrosion rate by a considerable amount.

KEY WORDS: corrosion, linear polarization, polarization resistance, carbon steel corrosion, ferritic stainless steel corrosion, geothermal corrosion, geothermal, oxygen corrosion effects, scaling

Geothermal brines contain significant quantities of dissolved solids in a reduced state, which can be very corrosive to the operating equipment. Significant amounts of dissolved chlorides, sulfides, silicates, carbon dioxide CO_2-carbonates, and others, as well as high liquid velocities and the presence of particulates, can result in a short equipment life. Conditions controlled by the operator, such as the entrance of oxygen, pH control by acid injection or control of CO_2, and the addition of scale inhibitors, can

[1] Senior research scientist, Battelle Pacific Northwest Laboratory, Richland, Wash. 99352.

also result in conditions leading to enhanced corrosion rates. A corrosion rate monitoring system is needed that will respond to rapidly changing corrosion conditions and that is easily available for installation by the plant operator. This study is concerned with determining whether commercial corrosion measuring equipment using the polarization resistance (PR) method can be easily adapted for use in the geothermal environment.

There have recently been two notable reviews on the general subject of the PR technique by Mansfeld [1][2] and Callow et al [2]. The Callow paper includes a useful summary of polarization resistance data on many alloys (such as iron, aluminum, and copper-nickel alloys). In general, the technique is considered able to predict corrosion rates in the laboratory within a factor of 2. Out in the field, where the interplay of the corrosion rate determining variables is more complex, the ability to predict the actual corrosion rates may be poorer. The theoretical relationships for the PR technique are shown here

$$I_{corr} = B \cdot 1/R_p \tag{1}$$

where

$$B = \beta_a \beta_c / 2.30(\beta_a + \beta_c) \tag{2}$$

and

$$1/R_p = \Delta I / \Delta E \tag{3}$$

The corrosion current density, I_{corr}, is linearly related to the reciprocal of the polarization resistance, R_p. When the system is polarized a few millivolts from the corrosion potential ($\Delta E < 20$ mV), the measured current, ΔI, permits the calculation of R_p. The proportionality constant, B, is related to the anodic and cathodic Tafel slopes, β_a, β_c, respectively. Although the Tafel slopes vary widely, the B value is relatively constant and, for carbon steels and stainless steels [2], falls within the range of 10 to 45 and 18 to 41 mV, respectively. Commercial PR instrumentation is designed to make use of Eq 1 by polarizing the corroding specimen ±10 mV or ±20 mV, measuring the current, and directly reading out the corrosion rate in mils per year. Equation 1 applies only under the following conditions:

1. The mixed potential theory applies, and the corrosion potential of the system is >50 mV from the reversible potential of either the anodic or cathodic reactions [1].
2. Both the anodic and cathodic reactions are activation controlled.

[2] The italic numbers in brackets refer to the list of references appended to this paper.

Both these conditions are often found in the corroding environment. However, when the cathodic reaction is under mass transport control (such as when oxygen enters the system), Eq 1 no longer applies. The PR relationship under conditions when the corrosion rate is controlled by mass transport of the cathodic process is shown here

$$I_{corr} = (\beta_a/2.3)(1/R_p) \tag{4}$$

One of the unresolved problems with PR methods is the difficulty of recognizing whether Eq 1 or Eq 4 applies, since there can be a large difference between B and β_a (especially if the metal is partially passivated).

The problem of the linearity of $(\partial I/\partial E)$ near E_{corr} has been extensively examined in the PR literature [1]. This is because I_{corr} is proportional to $(\partial I/\partial E)_{E_{corr}}$ only at the corrosion potential, and any nonlinearity of $I = f(E)$ could lead to an error in $(\partial I/\partial E)_{E_{corr}}$ obtained by extrapolation from ± 10 or ± 20 mV. Laboratory [1] studies have revealed that $I = f(E)$ is often nonlinear, but that the error is not significant when E_{corr} is greater than RT/F (R = gas constant; T = Kelvin temperature scale; and F = Faraday's constant) from both the equilibrium potential of the corrosion reaction and the reduction reaction. Conditions for a small error are usually encountered.

The only paper on the use of commercial PR instrumentation in geothermal brines has recently been published by Harrar et al [3]. They carried their experiment out at Niland, Calif. (with a brine of 20 percent solids content), using Petrolite (Houston, Tex.) 3-electrode equipment. The brine was cooled down to 100°C for all corrosion measurements, and acid injection was used to lower the pH. In one experiment, the PR electrodes were used as weight loss coupons, but the exposure times were only 16 to 25 hs. The weight loss data for carbon steels and chromium-molybdenum steel did not correlate well with the PR data [PR data for carbon steel was 2.5 and 6.3 mm/year (100 and 250 mils per year) for pH 3.4 and 2.3, respectively], though the correlation was good for the 400-series alloys.

This paper evaluates the use of the PR technique at 150 and 250°C in 3 and 20 percent brines, with and without oxygen. In general, the agreement between weight loss coupons and the PR technique is good. Commercial instrumentation should be adequate to monitor corrosion rates in the geothermal environment under reducing conditions.

Procedure

All experiments were carried out in a refreshed 1-litre titanium autoclave. No drawing of the setup is shown, but the interested reader is directed to a similar setup by Posey et al [4]. The solutions were kept in a 55-gal polyethylene drum and injected into the autoclave at about 1 litre/h.

Deoxygenated conditions were maintained by a CO_2 purge plus the use of a small amount of hydrazine. The oxygen partial pressure (typically 1 ppm in solution) was controlled by mixing air and CO_2 and sparging through the solution. The oxygen in solution was measured by the Winkler method. The compositions of the solutions are shown in Table 1.

A reference electrode is shown in Fig. 1 which permits the potential to be placed on the high-temperature hydrogen scale. Further details are discussed elsewhere [5], but the electrode operates at autoclave pressure and ambient temperature. Consequently, the potential contains a contribution from a thermal liquid junction potential. The Soret effect is slow to develop, so the potential is relatively constant with time, and this permits the design to be referenced to the high-temperature hydrogen electrode. These data are shown in Table 2. No attempt was made to correct the potential for the liquid junction potential.

Figure 2 shows the mounting of the electrode. Cylinder electrode specimens were used to be compatible with the results attained from the cylindrical weight loss specimens used in earlier experiments [6]. Belleville spring washers were used to keep a constant pressure on the polytetrafluoroethylene (PTFE) seals to minimize crevice corrosion problems. The specimen electrode holder is shown in Fig. 3. It is an improved design by Agrawal and associates [7]. The holder is constructed of corrosion-resistant zirconium metal to minimize any stray corrosion currents, and it is covered with a heat-shrinkable PTFE to further minimize stray current losses. It was tested without the electrodes by inserting a PTFE threaded bolt in the end and heating it in the autoclave containing brine to 150°C. The a-c impedance between it and the autoclave was greater than 10^6 ohms. Electrode specimens of ASTM Type A53B carbon steel (0.27 percent carbon, 0.71 percent phosphorus, and the remainder iron) and E-Brite 26-1 steel (26.2 percent chromium, 1.0 percent molybdenum, <.001 percent carbon, and the remainder iron) of 6.7 cm^2 were included in each experiment. For deoxygenated experiments, there were five weight loss specimens of Type A53B and E-Brite, each. Oxygenated experiments used two weight loss specimens each to reduce the rate of oxygen consumption. The duration of each experiment was 7 to 8 days in order for the corrosion rates to approach a steady state.

Electrochemical measurements were made with a potentiostat. Slow potential scan-current measurements (0.02 mV/s) up to ±20 mV from E_{corr} were made each day and the $(\partial I/\partial E)_{E_{corr}}$ was determined. After the slow scan methods indicated that a steady corrosion rate was obtained, voltage step-function measurements at ±10 mV were made to simulate the operation of some commercial PR instruments. On the final day of the experiment, Tafel slopes were measured at the same slow scan rate (up to ±100 mV from E_{corr}), followed by cyclic voltammetry (CV) scans at 10 mV/s

TABLE 1—*Synthetic brine compositions.*

No Oxygen	Oxygen
Low Salinity	
3% NaCl	3% NaCl
450 ppm SiO$_2$ (as SiO$_3^{2-}$)	450 ppm SiO$_2$ (as SiO$_3^{2-}$)
0.5 ppm S^{2-}	700 ppm SO$_4^{2-}$
Hydrazine (5 drops of 64%/200 litres)	CO$_2$-air sparge
1 atm CO$_2$	pH = 5.7 at 25°C
pH = 5.7 at 25°C	
High Salinity	
20% NaCl	20% NaCl
450 ppm SiO$_2$ (as SiO$_3^{2-}$)	450 ppm SiO$_2$ (as SiO$_3^{2-}$)
0.5 ppm S^{2-}	700 ppm SO$_4^{2-}$
Hydrazine (5 drops of 64%/200 litres)	CO$_2$-air sparge
1 atm CO$_2$	pH = 5.5 at 25°C
pH = 5.5 at 25°C	

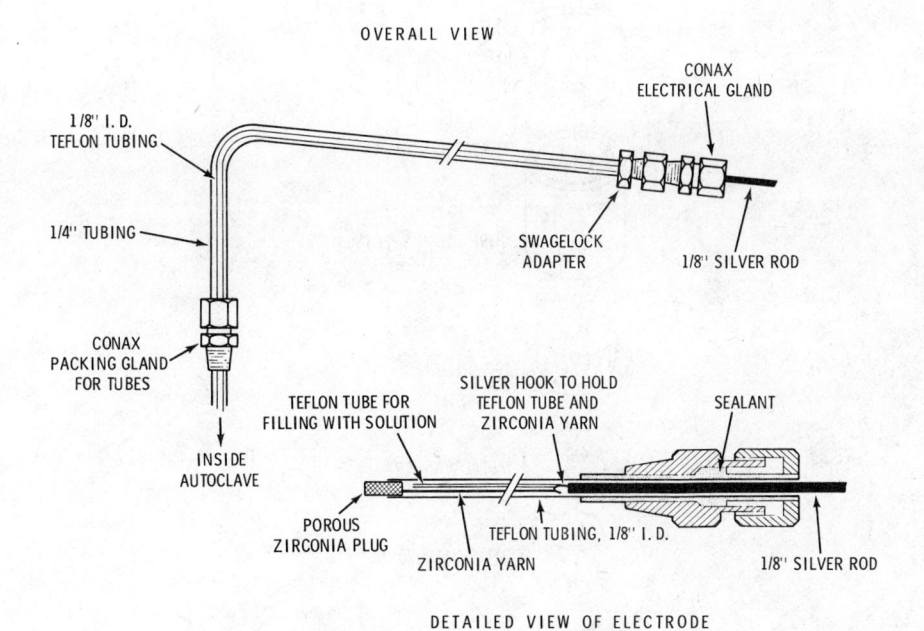

FIG. 1—*Detailed view of quaisi-reference electrode.*

TABLE 2—*Quaisi-reference electrode potentials (0.100m KCl) on the standard hydrogen scale.*

Temperature, deg C	$E_{Ag,AgCl_{25°C}} - E_{SHE_T}^0$
25	0.287 V
90	0.227 V
125	0.188 V
150	0.159 V
200	0.097 V
250	0.026 V

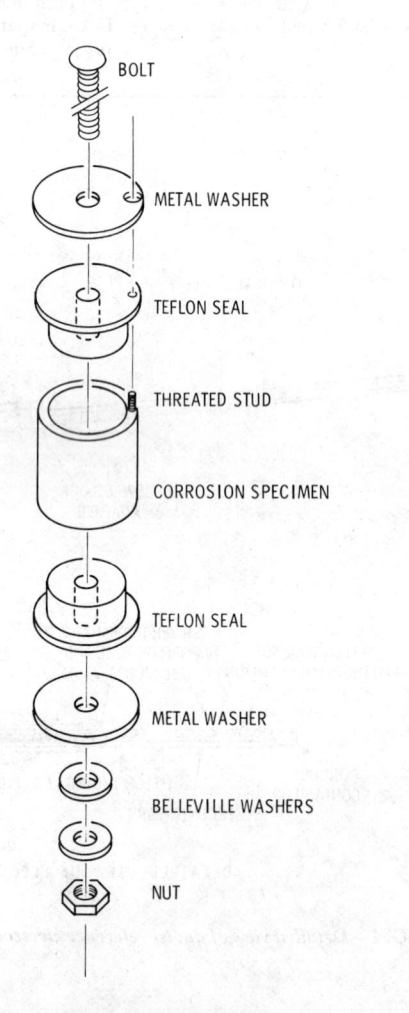

FIG. 2—*Electrode specimen.*

15.2 cm (6") LONG
0.33 cm (1/8") DIAMETER

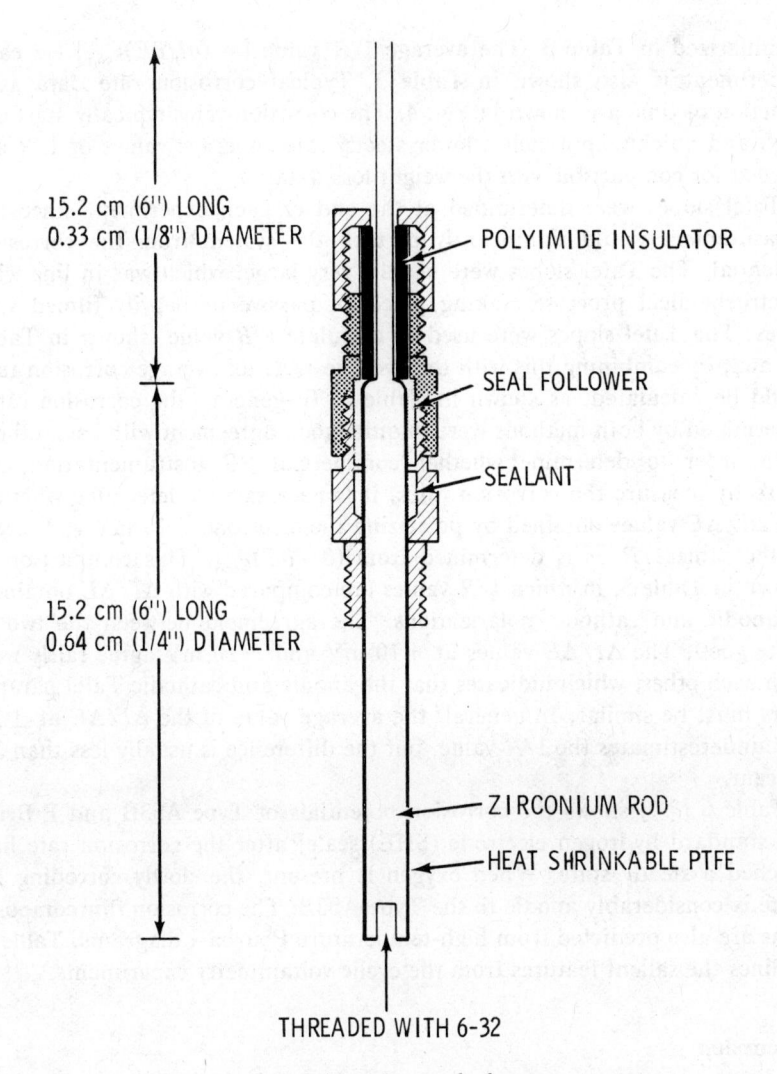

POLYIMIDE INSULATOR

SEAL FOLLOWER

SEALANT

15.2 cm (6") LONG
0.64 cm (1/4") DIAMETER

ZIRCONIUM ROD

HEAT SHRINKABLE PTFE

THREADED WITH 6-32

FIG. 3—*Electrode holder using standard pressure connector.*

at ±1 V extremes from E_{corr}. The CV experiments always started in a cathodic direction.

Results

The experiments were carried out at 150 and 250°C with and without oxygen. Only low-salinity brines were used at 150°C, but both high-salinity and low-salinity brines were used at 250°C. The results of these tests are summarized in Table 3. The average $1/R$ value [at $(\partial I/\partial E)_{E_{\text{corr}}}$] for each experiment is also shown in Table 3. Typical corrosion rate data as a function of time are shown in Fig. 4. The corrosion rates typically start out high and quickly approach a lower steady rate. Average values of $1/R$ are needed for comparison with the weight loss data.

Tafel slopes were determined at the end of each experiment since the measurement would significantly perturb the system from the corrosion potential. The Tafel slopes were usually very large, which was in line with electrochemical processes taking place on passive or heavily filmed surfaces. The Tafel slopes were used to calculate a B value (shown in Table 4), and, by combining this with the average $1/R$, an average corrosion rate could be calculated, as shown in Table 3. In general, the corrosion rates determined by both methods were in quite good agreement with each other.

In order to determine whether commercial PR instrumentation will correctly measure the corrosion rates, it is necessary to determine whether the $\Delta I/\Delta E$ values obtained by polarizing the electrode ± 10 mV will result in the same $1/R$ as is determined from $(\partial I/\partial E)_{E_{\text{corr}}}$. This comparison is shown in Table 5, in which $1/R$ values are compared with $\Delta I/\Delta E$ obtained at anodic and cathodic polarizations. The agreement between the two is quite good. The $\Delta I/\Delta E$ values at $+10$ mV and -10 mV agree fairly well with each other, which indicates that the anodic and cathodic Tafel parameters must be similar. In general, the average value of the $\Delta I/\Delta E$ at ± 10 mV underestimates the $1/R$ value, but the difference is usually less than 30 percent.

Table 6 [8,9] shows the corrosion potentials of Type A53B and E-Brite [on standard hydrogen electrode (SHE) scale] after the corrosion rate has reached a steady state. When oxygen is present, the slowly corroding E-Brite is considerably anodic to the Type A53B. The corrosion film compositions are also predicted from high-temperature Pourbaix diagrams. Table 7 outlines the salient features from the cyclic voltammetry experiments.

Discussion

The primary purpose of this work was to determine whether commercial PR instrumentation could correctly measure corrosion rates. The data in Table 3 indicate good agreement between weight loss measurements and the $1/R$ value determined from linear polarization experiments. Other experiments involving a ± 10 mV step function (to determine $\Delta I/\Delta E$), which replicate the operation of some commercial instruments, were in good agreement with the linear polarization data. In general, the agreement between weight loss and linear polarization data was considerably better than the factor of 2 agreement mentioned by others [1,2].

By using the weight loss data and the average $1/R$, a B value was calculated, as shown in Table 4. It was felt that this was a more accurate value

TABLE 3—Corrosion data.

	A53B				E-Brite		
Experiment	Weight Loss, mg/dm² day (mils per year)	Linear Polarization, mg/dm² (mils per year)	Average I/R, µA/mV		Weight Loss, mg/dm² (mils per year)	Linear Polarization, mg/dm² (mils per year)	Average I/R, µA/mV	Notes
1.0	26.3 ± 1.1 (4.8 ± 0.2)	19.2 (3.5)	2.4	2.52 ± 0.11 (0.46 ± 0.02)	...	0.37	low salinity, no O$_2$, 150°C
1.1	138.6 (25.3)	...	2.1	15.3 (2.8)	...	0.90	low salinity, 1 ppm O$_2$, 150°C
1.2	12.11 ± 0.11 (2.21 ± 0.02)	20.3 (3.7)	1.5	3.95 ± 0.27 (0.72 ± 0.05)	7.12 (1.3)	0.47	low salinity, no O$_2$, 250°C
1.3	221.9 ± 18.1 (40.5 ± 3.3)	197 (36)[a]	4.6	3.23 ± 0.22 (0.59 ± 0.04)	5.48 (1.0)	0.41	low salinity, 1 ppm O$_2$, 250°C
1.4	40.8 ± 2.6 (7.45 ± 0.47)	35.6 (6.5)	4.7	3.67 ± 0.27 (0.67 ± 0.05)	7.12 (1.3)	0.46	high salinity, no O$_2$, 250°C
1.5	41.6 ± 3.3 (7.6 ± 0.6)	15.9 (2.9)	4.7	1.92 ± 0.05 (0.35 ± 0.01)	2.69 (0.49)	0.14	high salinity, 1 ppm O$_2$, 250°C

[a] The mass transport equation was used for calculation.

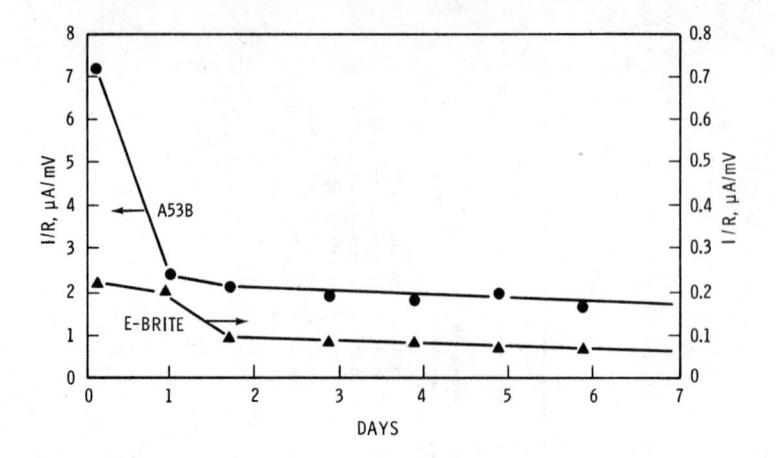

FIG. 4—*Typical polarization resistance data (Experiment 1.0 shown) as a function of time.*

than that calculated from Tafel slopes because it was based on actual corrosion data. For Type A53B in the absence of oxygen, B was 24 ±3 mV, whereas, for all cases of E-Brite, B was 28 ±11 mV. Commercial instruments have this B value hardwired in place, and it is important to know this value if it becomes necessary to correct the commercial instrument output. The B value for the Petrolite (Houston, Tex.) PR instrument is calculated to be 36 mV, based on the electrode area for their standard Type 1018 mild steel electrode. The internal B value for the Corrater (Magma instruments, Santa Fe Springs, Calif.) is not known. Based on the variability of B values measured in the laboratory (especially for E-Brite) and the fact that corrosion conditions in the field are more variable and different from the laboratory conditions, the hardwired B value of 36 mV in the Petrolite is acceptable as a direct measure of the corrosion rates, since the error is less than a factor of 2 when compared with the average B values.

When oxygen enters the system, the cathodic reaction approaches mass transport control. Equation 4 must be used to calculate the corrosion rate. Unfortunately, the $\Delta I/\Delta E$ behavior (from commercial instruments) does not give any warning that the corrosion conditions have changed. Examination of the $1/R$ data of Experiments 1.1 and 1.3 in Table 3 does not show any large change in the $1/R$ data; consequently, by assuming that Eq 1 and 2 always hold, the actual corrosion rate could be underestimated by a factor of 5 or more. The only clue is the potential data of Table 6. By mixed potential theory, the corrosion potential will always be closest to the reaction with the largest exchange current density. The Type A53B electrode is corroding relatively freely, and, when the rapidly reacting oxygen

TABLE 4—*Tafel data*.

Experiment	Anodic Tafel A53B, mV	Anodic Tafel E-Brite, mV	Cathodic Tafel A53B, mV	Cathodic Tafel E-Brite, mV	B, From Tafel Data A53B, mV	B, From Tafel Data E-Brite, mV	B, From Corrosion Data A53B, mV	B, From Corrosion Data E-Brite, mV	Notes
1.0	158	...	72	...	21.5	...	29	18	low salinity, no O_2, 150°C
1.1	...	...	...	...	...	...	177	46	low salinity, 1 ppm O_2, 150°C
1.2	378	234	108	152	36	40	22	22	low salinity, no O_2, 250°C
1.3	265	349	182	114	65	30	129	21	low salinity, 1 ppm O_2, 250°C
1.4	191	534	61	115	20	41	23	21	high salinity, 1 ppm O_2, 250°C
1.5	26	196	120	299	9	51	24	37	high salinity, 1 ppm O_2, 250°C

TABLE 5—Polarization resistance data.

Experiment	A53B $(\partial I/\partial E)_{E_{corr}}$, $\mu A/mV$	A53B $\Delta I/\Delta E$ +10 mV, $\mu A/mV$	A53B $\Delta I/\Delta E$ −10 mV, $\mu A/mV$	E-Brite $(\partial I/\partial E)_{E_{corr}}$, $\mu A/mV$	E-Brite $\Delta I/\Delta E$ +10 mV, $\mu A/mV$	E-Brite $\Delta I/\Delta E$ −10 mV, $\mu A/mV$
1.0	1.9	1.8	1.8	0.08	0.15	0.15
1.1	2.4	1.5	1.0	1.0	0.80	0.70
1.2	1.5	1.0	1.0	0.30	0.38	0.30
1.3	4.2	3.0	...	0.42	0.45	...
1.4	4.9	3.7	5.0	0.50	0.30	0.44
1.5	4.2	3.0	4.0	0.12	0.15	0.10

TABLE 6—*Steady-state corrosion potentials (on SHE).*

Experiment	A53B	E-Brite	Predicted Phases from Pourbaix Diagrams A53B	Predicted Phases from Pourbaix Diagrams E-Brite	Notes
1.0	−0.522	−0.264	FeS	FeS, Cr_2O_3	low salinity, no O_2, 150°C
1.1	−0.409	−0.106	Fe_3O_4	Fe_3O_4, Fe_2O_3, Cr_2O_3	low salinity, 1 ppm O_2, 150°C
1.2	−0.684	−0.646	FeS	FeS, Cr_2O_3	low salinity, no O_2, 250°C
1.3	−0.486	−0.229	Fe_3O_4	Fe_2O_3, Cr_2O_3	low salinity, 1 ppm O_2, 250°C
1.4	−0.599	−0.592	FeS, Fe_3O_4	Fe_3O_4, Cr_2O_3	high salinity, no O_2, 250°C
1.5	−0.571	−0.227	Fe_3O_4	Fe_2O_3, Cr_2O_3	high salinity, 1 ppm O_2, 250°C

TABLE 7—*Cyclic voltammetry data.*

Experiment		Observations, Cathodic Direction	Observations, Anodic Direction	Notes
1.0	A53B	H_2 evolution	general corrosion	low salinity, no O_2, 150°C
	E-Brite	H_2 evolution	passivity $E_{prot} - E_{corr} = 20$ mV	
1.1	A53B	H_2 evolution	general corrosion	low salinity, 1 ppm O_2, 150°C
	E-Brite	H_2 evolution	passivity $E_{prot} - E_{corr} = -150$ mV	
1.2	A53B	H_2 evolution	passive region	low salinity, no O_2, 250°C
	E-Brite	H_2 evolution	passivity $E_{prot} - E_{corr} = 240$ mV	
1.3	A53B	H_2 evolution	passive region	low salinity, 1 ppm O_2, 250°C
	E-Brite	H_2 evolution	passivity $E_{prot} - E_{corr} = 160$ mV	
1.4	A53B	H_2 evolution	general corrosion	high salinity, no O_2, 250°C
	E-Brite	H_2 evolution	passivity $E_{prot} - E_{corr} = 160$ mV	
1.5	A53B	film reduction peak and H_2 evolution	general corrosion	high salinity, 1 ppm O_2 250°C
	E-Brite	H_2 evolution	passivity $E_{prot} - E_{corr} = 170$ mV	

enters the system, the corrosion potential does not greatly change in the anodic direction. However, the E-Brite is passivated (small exchange current density), and the corrosion potential is shifted in a more anodic direction. This is particularly true at 250°C, when the rate of oxygen consumption should be faster. A thermodynamic reference electrode is required for this observation, but the commercial instruments use freely corroding specimens of the same material, so this observation could not be made. Either a thermodynamic reference electrode should be used, or the instruments should include a very slowly corroding specimen along with the iron electrode, and the potential between the slow and more rapidly corroding electrodes should be monitored to warn when oxygen has entered. When oxygen is known to be present, the reading of the commerciation PR instrumentation for steels should be multiplied by at least a factor 5 to arrive at a more realistic corrosion rate. Though only one oxygen level was studied in these experiments, mass transport control implies that the corrosion rates for a carbon steel will be proportional to the oxygen level.

Since the Tafel slopes are rarely the same for the anodic and cathodic reaction, the current measured at ±10 mV will not be the same as that at −10 mV, and the commercial instrument will give differing corrosion rates. A simple average of these two corrosion rates is more likely to approximate the $(\partial I/\partial E)_{E_{corr}}$ than either value separately and is a useful "rule of thumb" for the field personnel. This is discussed in more detail elsewhere [12].

The corrosion rates for Type A53B steel in high-saline brine are much lower than those observed by others [3,6]. This is believed to be due to the higher brine pH used in these experiments. Corrosion rates typically increase as the pH is decreased. As a first approximation, a first-order dependence of corrosion rate on hydrogen ion activity is possible. Harrar et al [3] studied corrosion rates in the pH range of 2.3 to 3.4 at 100°C and measured rates greater than 2.5 mm/year (100 mils per year) for carbon steel. Shannon [6] studied corrosion rates in 20 percent brine at 250°C (pH = 4.7) and measured rates on mild steel of about 2.5 mm/year (100 mils per year). Cramer et al [10] measured corrosion rates in 20 percent brine with a pH of 6.1 (simulated Salton Sea brine) at 232°C and found the general corrosion rate on mild steel to be 0.13 mm/year (5 mils per year) whereas, in 100 ppm O_2 brine, the corrosion rate was over 25 mm/year (1000 mils per year). The pH has a profound effect on corrosion rates, and the corrosion rates measured in this study are consistent with those measured by others.

The cyclic voltammetry (CV) data are shown in Table 7. Cyclic voltammetry scan rates are too rapid to approach a steady state, but the method can point out some general features of the current-potential relationship. For Type A53B steel, the anodic scan almost always indicated general corrosion, whereas, for E-Brite, passivation was generally present. Cyclic voltammetry can indicate in a general way the onset of pitting [11]. A material susceptible to pitting shows a larger current on the return half of the anodic scan than on the first half. The point at which the two currents are equal is defined as E_{prot}, the protection potential from pitting. Type A53B did not show any E_{prot}, that is, the current on the return half of the cycle was always below that on the first half of the cycle. The E-Brite always showed an E_{prot}, which was usually less than 200 mV from the corrosion potential. In Experiment 1.1, it was below E_{corr}, which indicated that pitting was taking place rapidly. Visual observation of the weight loss coupons never revealed pitting on the Type A53B steel, but the E-Brite was usually pitted even without the presence of oxygen. The E_{prot} was close to E_{corr}, so that occasional potential perturbations resulted in the initiation of pits. The pit density was rarely above 5 pits/cm^2 (except on the electrodes, which were made extremely anodic on CV scans) with a depth of less than 0.013 cm (0.005 in.). The passive condition of E-Brite explains the resistance to uniform corrosion of this alloy, but the pitting makes it unsuitable for engineering application in brines. The general anodic corrosion behavior of the Type A53B steel, based on CV observation, indicates that the addition of depolarizing species (such as O_2) and improved mass transport conditions will result in increased corrosion rates. Cathodic behavior for both materials did not reveal any current peaks, and the return cycle (second half of the cathodic scan) generally was superimposed on the first cycle. Experiment 1.5 was an exception in the case of Type A53B. A reduction peak was observed on the first half of the cathodic cycle but was not seen on later CV scans. Experiment 1.5 is interesting because the Type A53B steel did not show an enhanced corrosion rate in the presence of oxygen, though all previous experiments in low-salinity brines with oxygen resulted in enhanced corrosion rates. It appears that a protective film may be present, although the anodic CV scan did not show any passivation. Experiment 1.5 will be further investigated in the future.

There were no apparent surface films on E-Brite steel for any of these experiments. Type A53B typically had a black, porous, poorly adherent film under both oxygenated and reducing conditions, except in Experiment 1.5, when the film was black, quite thin, and adherent. X-ray diffraction studies were carried out on Type A53B surfaces exposed to reducing conditions at 150°C. Iron carbonate ($FeCO_3$) and iron sulfides (FeS and FeS_2) were found. From the data in Table 6, only FeS was predicted. No other surface determination studies were carried out at the higher temperatures. No Pourbaix diagram for elevated temperatures calculated to date considers the $FeCO_3$ species, which should be included for completeness. Work by Shannon [6] indicates that the corrosion rate (at a constant brine composition without sulfide) is much higher in the range of 50 to 150°C than above 150°C. Preliminary calculations of the stability field for $FeCO_3$ indicate that it is a dominant iron film at lower temperatures but is displaced by a more protective Fe_3O_4 at temperatures above 150°C.

Conclusions

The agreement between weight loss corrosion rates and those calculated from the polarization resistance technique is quite good. Examination of the operation of commercial polarization resistance instrumentation indicates that they should perform well in geothermal brines at elevated temperature. However, when oxygen or some other depolarizer enters the system, commercial instrumentation may give erroneously low readings.

Acknowledgments

The author wishes to thank D. W. Shannon (program manager) and S. J. Thompson (assistant during the early experiments) for their assistance. The author thanks Robert Reeber (U.S. Department of Energy, Geothermal Energy Division) for financial support. This work was performed for the U.S. Department of Energy under Contract EY-76C-06-1830.

References

[1] Fontana, M. and Staehle, R., Eds., *Advances in Corrosion Science and Technology,* Vol. 6, Plenum Press, New York, 1970, pp. 163-262.
[2] Callow, L., Richardson, J., and Dawson, J., *British Corrosion Journal,* Vol. 11, No. 3, 1976, pp. 123-139.
[3] Harrar, J., McCright, R., and Goldberg, A., "Field Electrochemical Measurements of Corrosion Characteristics of Materials in Hypersaline Geothermal Brines," Report UCRL-52376, Lawrence Livermore Laboratory, Livermore, Calif., 1977.
[4] Posey, F. and Palko, A., *Corrosion,* Vol. 35, No. 1, 1979, pp. 38-42.
[5] Danielson, M., *Corrosion,* Vol. 35, No. 5, 1979, pp. 200-204.
[6] Shannon, D., "Corrosion of Iron-Base Alloys Versus Alternative Materials in Geothermal Brines," Report PNL-2456, Battelle Pacific Northwest Laboratories, Richland, Wash., 1977.
[7] Agrawal, A., Damin, D., McCright, R., and Staehle, R., *Corrosion,* Vol. 31, 1975, p. 262.
[8] Syrett, B., MacDonald, D., Shih, H., and Wing, S., "Corrosion Chemistry of Geothermal Brines, Part 2," Report AER76-00713, Stanford Research Institute, Menlo Park, Calif., 1977.
[9] Biernat, R. and Robins, R., *Electrochimica Acta,* Vol. 17, 1972, pp. 1261-1283.
[10] Cramer, S., Carter, J., McCawley, F., and Needham, P., "Corrosion Studies in High Temperature, Hypersaline Geothermal Brines," Paper No. 59, National Association of Corrosion Engineers, *Corrosion/79,* Atlanta, Ga., 1979.
[11] Baboian, R., Ed., *Electrochemical Techniques for Corrosion,* National Association of Corrosion Engineers, Houston, Tex., 1977.
[12] Danielson, M. J., *Corrosion,* Vol. 36, No. 4, 1980, pp. 174-177.

G. H. Schnaper,[1] *V. R. Koch,*[2] *and S. B. Brummer*[3]

Corrosion Protection of Solar-Collector Heat Exchangers and Geothermal Systems by Electrodeposited Organic Films

REFERENCE: Schnaper, G. H., Koch, V. R., and Brummer, S. B., "**Corrosion Protection of Solar-Collector Heat Exchangers and Geothermal Systems by Electrodeposited Organic Films,**" *Geothermal Scaling and Corrosion, ASTM STP 717,* L. A. Casper and T. R. Pinchback, Eds., American Society for Testing and Materials, 1980, pp. 57–68.

ABSTRACT: Solar-collector heat exchangers normally experience temperatures up to 150°C (300°F), and stagnation temperatures may reach 200°C (400°F). Geothermal systems, in turn, often operate at temperatures near 250°C (480°F). Despite the presence of inhibitors, these high temperatures increase the corrosion rate of the metal to intolerable levels, thus reducing the lifetime of the system. We have found that by electrochemically depositing thin polymeric films on typical solar-collector and geothermal system materials (copper, iron, and aluminum), the corrosion rates can be substantially reduced. In particular, a 4-(p-hydroxyphenyl)-2-butanone polyphenylene oxide film subsequently treated with 2,4-dinitrophenylhydrazine on a mild steel substrate afforded minimal corrosion currents.

KEY WORDS: corrosion, geothermal systems, solar collector systems, scaling, organic films, electropolymerization

Under normal operating conditions, solar-collector heat exchangers can experience temperatures of up to 150°C (300°F); moreover, stagnation temperatures may exceed 200°C (400°F) [1].[4] Geothermal systems, in turn, often operate at temperatures near 250°C (480°F). Despite the presence of inhibitors, these high temperatures increase the corrosion rate of the metals to intolerable levels, thus reducing the lifetime of the system.

Electrocoatings have been used since the 1960s to protect metals from

[1] Staff scientist, EIC Corp., Newton, Mass. 02158.
[2] Group leader, EIC Corp., Newton, Mass. 02158.
[3] Vice president, EIC Corp., Newton, Mass. 02158.
[4] The italic numbers in brackets refer to the list of references appended to this paper.

corrosion [2]. However, these coating processes involve the electrophoretic deposition of macromolecules (paint) of a previously formed polymer. By placing the metal in a solution of the monomer and forming the polymer directly on the metal surface, several benefits can be obtained:

1. The absence of a dissolved inhibitor permits operation with untreated water and cuts capital and operating costs.
2. The process may allow the use of otherwise corrodible, inexpensive materials, such as mild steel.
3. The films may be deposited in very thin coherent layers on the surfaces of objects with complex shapes.
4. The films should be intrinsically pinhole free, but, should they develop faults, they can be repaired *in situ*.

We have, therefore, focused on minimizing corrosion by the electroinitiated polymerization of a dissolved organic monomer onto a metal surface. This technique involves the electrooxidation of a suitable monomer at a fixed potential until the polymer film passivates the surface, as has been demonstrated previously [3-5].

Experimental

Metal Electrodes—Substrate metals evaluated include mild steel, copper, and aluminum. Rectangular coupons (25 by 38 mm), cut from 0.25-mm (10-mil) shim stock (Roblinger Division of Slavin Co., Boston), were degreased with methylene chloride prior to use. In some cases, coupons were etched in acid baths or mechanically abraded to remove surface films. Analytical disk electrodes of reproducible area were fabricated from 3-mm diameter stock, press-fit into polypropylene, and polished to a mirror surface with metallographic grade alumina (Fisher).

Chemicals—The organic monomers employed in this study were: (I) (*m*-phenylenediamine (MPD), (II) 2,6-dimethylphenol (DMP), (III) 2,3,5,6-tetramethylphenol (TMP), (IV) *m*-aminophenol (MAP), and (V) 4-(*p*-hydroxyphenyl)-2-butanone (PHPB). All were obtained from Aldrich and used without further purification.

Other reagents included 2,4-dinitrophenylhydrazine (DNPH), *p*-phenylenediamine (PPD), and terephthalaldehyde (TPA) (Aldrich).

Electrochemical Cell and Instrumentation—A three-electrode, two-compartment cell (Fig. 1) was used for cyclic voltammograms (CV) and the electrodeposition of films onto metal coupons. All potentials are quoted in relation to a saturated calomel electrode (SCE) separated from the central compartment by a fine glass frit. The electrolyte was typically 0.2 *M* sodium methoxide in methanol (NaOMe/MeOH), although some electroanalytical work utilized aqueous solutions buffered to pH 9.18. The electrolyte was always deaerated for 15 min with nitrogen (N_2) before any

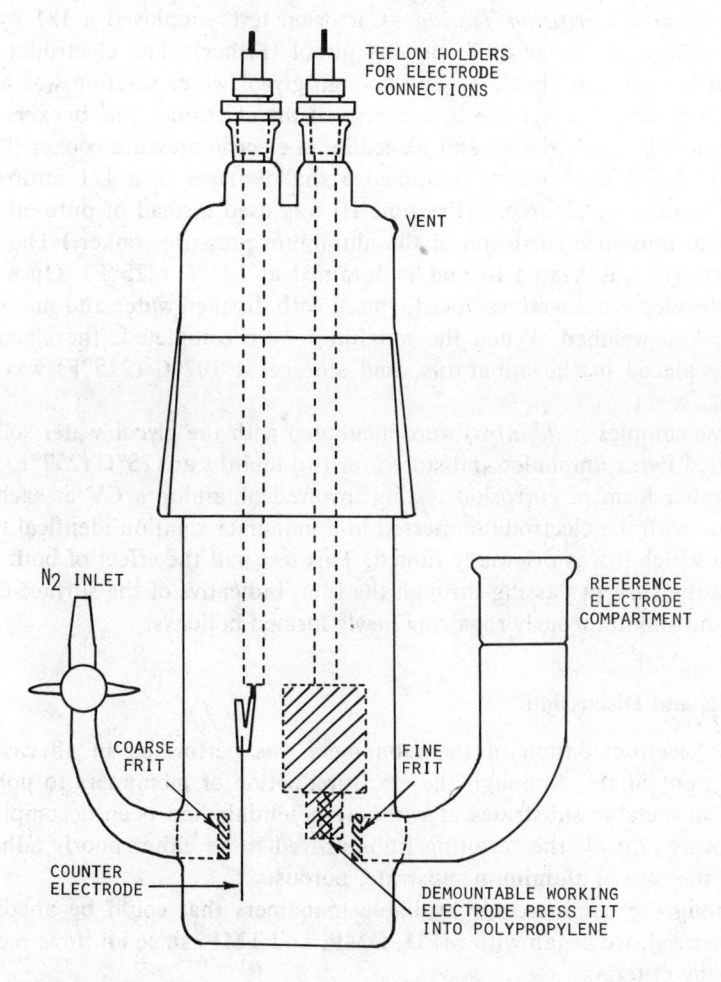

FIG. 1—*Two-compartment electrochemical cell for cyclic voltammetry and deposition of electrogenerated polymer film.*

electrochemistry was done. The working electrode consisted of the metal coupon or analytical disk electrode undergoing polymerization. The counter electrode was a 20-mm^2 section of platinum foil.

Anodic potentials were applied for a period of 1 hour via a Brinkmann Model LT73 potentiostat. The current-time curve was monitored throughout on a Houston Omniscribe 2000 x-y recorder equipped with a time base. The electrolysis was performed at ambient temperature under an N_2 atmosphere to prevent auto-oxidation of the monomer. After the deposition, the electrode was rinsed with methanol and distilled water and allowed to air-dry. Electron spectroscopy for chemical analysis (ESCA) of filmed coupons was performed by Surface Science Laboratories (Palo Alto, Calif.).

Accelerated Corrosion Testing—Corrosion tests employed a 1/1 by volume solution of water and ethylene glycol (Fisher). The electrodes were placed in individual beakers, and enough glycol/water solution was added to each beaker to cover the filmed area of the electrode. The beakers were covered with watch glasses and placed in an electric pressure cooker (Presto Model 02/PCE6H), which contained a shallow layer of a 1/1 antifreeze/water solution. (Antifreeze [Prestone II] was used instead of pure ethylene glycol to minimize corrosion of the aluminum pressure cooker.) The pressure cooker was heated to and maintained at 107°C (225°F). Once each day, the electrodes were removed, rinsed with distilled water and methanol, dried, and weighed. When the weighings were completed, the electrodes were replaced in the apparatus, and storage at 107°C (225°F) was continued.

Some samples (*vide infra*) were incubated with the glycol/water solution in sealed Pyrex ampoules and stored for two months at 125°C (257°F).

Another form of corrosion testing involved obtaining a CV at each test interval with the electrode immersed in a monomer solution identical to the one in which it was originally filmed. This test had the effect of both measuring the current passing through the film, indicative of the surface coverage, and simultaneously repairing newly formed holidays.

Results and Discussion

The electrooxidation of the monomers was performed in all cases at anodic potentials. Although the electroreduction of monomers to polymer films on metallic substrates at cathodic potentials had been accomplished previously [3,6,7], the resulting films proved to be either poorly adherent or, in the case of aluminum substrate, porous.

Among the commercially available monomers that could be anodically polymerized, we began with MPD, DMP, and TMP, since all three met the following criteria:

(*a*) electroactivity in the potential range of 0.0 to 1.00 V versus SCE;

(*b*) polymerization, rather than dimerization or long-term stability of electron-deficient intermediates;
(*c*) adhesion to the metal surface after polymerization; and
(*d*) good solubility in our solvents of choice: water and methanol.

DMP and TMP have the added advantage of allowing polymerization only at the *para* position of the aromatic ring because of the blocking methyl substituents. This leads to the formation of a regular polyphenylene oxide (PPO) film through a free-radical mechanism as has been suggested by Pham and coworkers [*4*] (Eqs 1 through 3).

$$\text{DMP} + \bar{\text{OH}} \xrightarrow{\text{MeOH}} \text{VI} \quad (1)$$

$$\text{VI} \xrightarrow{e^-} \text{VII} \rightleftharpoons \text{VIII} \quad (2)$$

$$\text{VII} + \text{VIII} \longrightarrow \xrightarrow{-H^+} \text{IX} \; \text{etc.} \quad (3)$$

Prior to deposition on coupons, CVs of the monomer were obtained on glassy carbon and on the analytical disk electrodes in order to determine the proper deposition potential. Figure 2 shows typical CV scans of the electrolyte with and without the monomer. The lower curve reveals little current over the entire potential range. The upper curve, taken after the addition of 2 mM MPD, manifests typical film-forming behavior. The first sweep in the anodic direction reveals a wave at +0.64 V, which indicates the oxidation of MPD at the electrode surface. Scan 2 depicts the second sweep, taken immediately after the first. As can be seen, the current has dropped off dramatically, and the oxidation wave has virtually

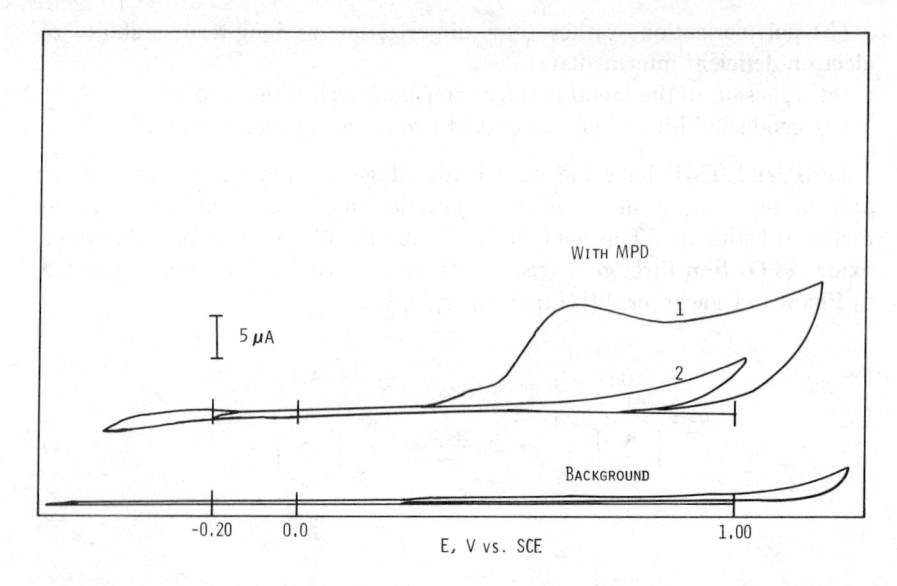

FIG. 2—*Cyclic voltammograms of the background electrolyte and the MPD monomer in pH 9.18 aqueous buffer.*

disappeared. Subsequent sweeps showed much lower currents, indicating electrode passivation. This passivation of the electrode surface is a result of the formation of polymer film. The reaction takes place at the surface, and, because of adsorption onto the surface, the film does not diffuse away. As the polymer film builds up on the metallic surface, monomer oxidation ceases because the film acts as an insulating barrier, preventing the passage of additional current. Ideally, complete coverage would be indicated by complete passivation, or zero current. Edge effects and other surface phenomena prevent this; however, we have obtained residual currents of less than 0.1 $\mu A/cm^2$ during long-term (24-h) deposition.

Films have been deposited on all three substrate metals. In order to obtain a film on aluminum, we have found it necessary to replace the surface oxide with zinc, through the zinc immersion process [8]. The polymer, therefore, is actually formed on a zinc surface rather than on aluminum oxide. Even with this preparation, the film appeared to be very incomplete. This was borne out by the ESCA analysis (*vide infra*).

Cross-Linked Polymers—Early studies showed TMP to be the most suitable PPO film precursor in terms of electrode passivation and physical appearance. However, further examination proved this PPO film to be poorly adherent and easily removed by a Scotch tape yank test used by Pham and coworkers [5]. They also found that by incorporating a carbonyl substituent (such as an aldehyde or ketone) into the monomer and reacting the electrogenerated PPO film with 2,4-dinitrophenylhydrazine (DNPH), a poly-(Schiff base) is formed [5] (Eq 4).

$$\left(Ar-\overset{R}{\underset{||}{C}}=O\right)_n + H_2NHN-\underset{NO_2}{\underset{DNPH}{\bigcirc}}-NO_2 \xrightarrow{H^+/EtOH} \left(Ar\overset{R}{\underset{|}{C}}=NNH-\underset{NO_2}{\bigcirc}-NO_2\right)_n + nH_2O \quad (4)$$

The degree of surface coverage was increased and the thermal stability of the polymer improved by this reaction.

Our idea for improving the physical and chemical properties of the polymer film involved reacting a diamine or dialdehyde, as in Eq 5

$$2\left(Ar-\overset{H}{\underset{|}{C}}=O\right)_n + H_2N-\bigcirc-NH_2 \longrightarrow \left(Ar-\overset{H}{\underset{|}{C}}=N-\bigcirc-N=\overset{H}{\underset{|}{C}}-Ar\right)_n + 2nH_2O \quad (5)$$

PPD

with an appropriately substituted PPO film to cross-link the polymer chains. This heterogeneous reaction of the surface film with the cross-linking agent in solution should produce better surface coverage, greater thermal stability, and stronger adherent qualities to the film. Accordingly, several new monomers incorporating carbonyl or amino substituents, including m-aminophenol and 4-(p-hydroxypheny)-2-butanone, were electrochemically polymerized.

ESCA Results—Table 1 shows the results of ESCA studies on some of these films. One immediately notes that the substrate metal can be detected through all the films examined by the ESCA technique. One possible explanation is that the films are extremely thin, and the scanning X-ray beam penetrates through to the substrate. This type of film would be similar to the Class III cathodic films obtained by Subramanian and co-workers [9]. Another explanation is that the films are sufficiently thick, but contain holidays, such as pinholes or cracks, through which the substrate can be detected.

The last two columns are the calculated and observed carbon/heteroatom ratios for the proposed structures of the polymer films. Most of the observed values are lower than the calculated values. It is believed that this is due to unfilmed substrate, which is, in actuality, a metal oxide layer. Thus, the additional oxygen lowers the carbon/heteroatom ratio. For the MPD film on copper, however, the observed value is greater than the calculated value. Also, the detected substrate is much higher and the nitrogen much lower than those for the same film on steel. We suspect that some

TABLE 1—*Results of ESCA studies on electropolymerized films.*[a]

Film/Substrate	C	O	N	Cu	Fe	Zn	Calculated C/(O + N)	Observed C/(O + N)
TMP/Fe	89	11	...[b]	...	0.2	...	10	8.1
TMP/Cu	86	12	1.7	0.2	...	...	10	6.3
TMP/Al-Zn	59	28	1.4	...	...	12	10	2.0
MPD/Fe	73	14	11.0	...	0.3	...	3	2.9
MPD/Cu	79	14	4.3	0.8	...	...	3	4.3
MPD-TPA/Fe	74	17	6.1	0.1	0.2	...	5	3.2
MPD-TPA/Cu	62	21	3.5	0.5	...	...	5	2.5

[a]The values are expressed as atom percent from integrated peak areas.
[b]<0.1 atom percent detected.

chemical reaction is occurring at the electrode surface, such as oxidation or hydrolysis, which depletes the nitrogen content and creates or enlarges "holes" in the film.

MPD-TPA is an MPD film that has been reacted heterogeneously with terephthalaldehyde (X).

CHO–C₆H₄–CHO

X

Comparison of the detected substrate would indicate that the cross-linked film is thicker and therefore has better surface coverage than the noncross-linked film. The film also changes color slightly upon reaction with the TPA solution, which is further evidence that cross-linking has occurred. However, the carbon/heteroatom ratios show the opposite trend from what was expected. Our data, therefore, are somewhat ambiguous.

Electron spectroscopy for chemical analysis for a TMP film deposited onto zinc-plated aluminum is also shown in Table 1. The visible irregularities in the film are corroborated by the large increase in detected substrate and the large decrease in detected carbon. The zinc immersion process was therefore deemed unsuitable for aluminum substrates.

Corrosion Tests

Weight Loss Method—Figure 3 shows a plot of total weight loss versus time of storage for TMP films on steel coupons. The plot can be divided into three sections of about 6 days each. In the first section (Days 0 through 6), the weight losses are approximately the same for all three coupons. In the second section (Days 6 through 12), the filmed coupons lost significantly less weight than the unfilmed control, and, in the third section (after Day 12), the weight losses for the filmed coupons accelerated to meet and surpass those of the control.

Since TMP films were visibly poorly adherent, we hypothesized a porous film, which would incorporate microscopic holes through to the surface. At the beginning of the test, the solvent worked directly on the metal surfaces through the holes, yielding similar weight losses. After several days, the holes became plugged with corrosion products, which minimized further weight loss, although the trapped solvent continued to corrode the metal. Finally, the film was undermined by the corrosion and lifted off the metal,

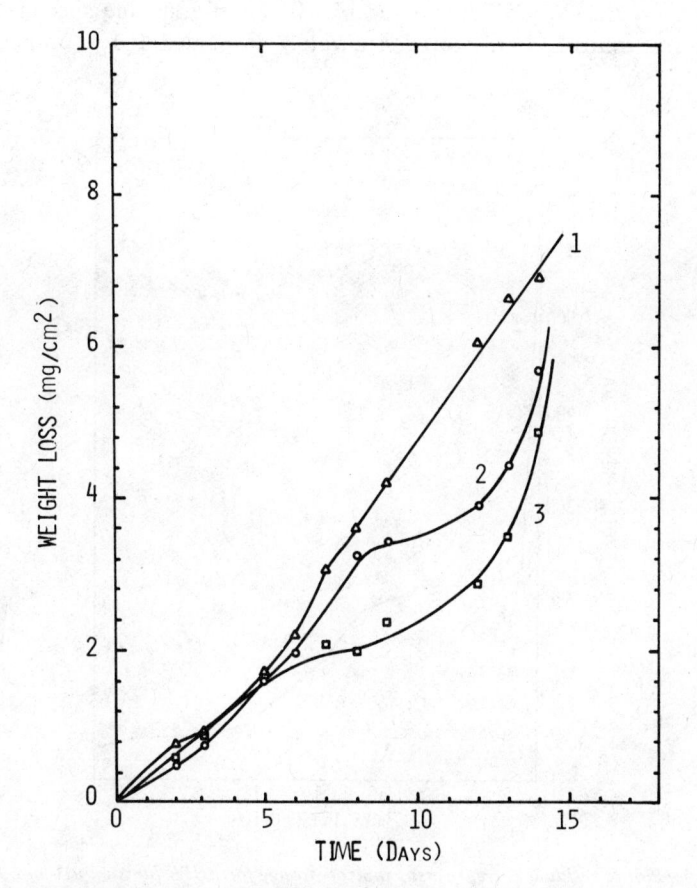

FIG. 3—*Plot of total weight loss versus time of storage for TMP films on steel coupons: (1) unfilmed control; (2) filmed, chemically etched prior to filming; (3) filmed, unetched.*

exposing fresh surface to direct corrosion and yielding the accelerated weight loss. This behavior is similar to that caused by filiform and crevice corrosion [10].

The film on the etched coupon gave less corrosion protection than that on the unetched, we believe, because the irregular surface contained insoluble products, left from the etch, that inhibit the adsorption of the polymer onto the surface and result in the formation of an incomplete film.

CV Method—Figure 4 is a plot of residual current versus time of storage for PHPB films that had been reacted with DNPH. It can be seen that the residual current falls off rapidly after the original filming, and then more slowly as the test progresses. Repeated CVs apparently fill the holes left by the corrosion testing and by the original polymerization. As more holidays are filled, less current passes through the system, and the residual current decreases.

Storage at 125°C (257°F)—Several TMP/aluminum-zinc specimens and aluminum controls were stored separately in sealed Pyrex ampoules at

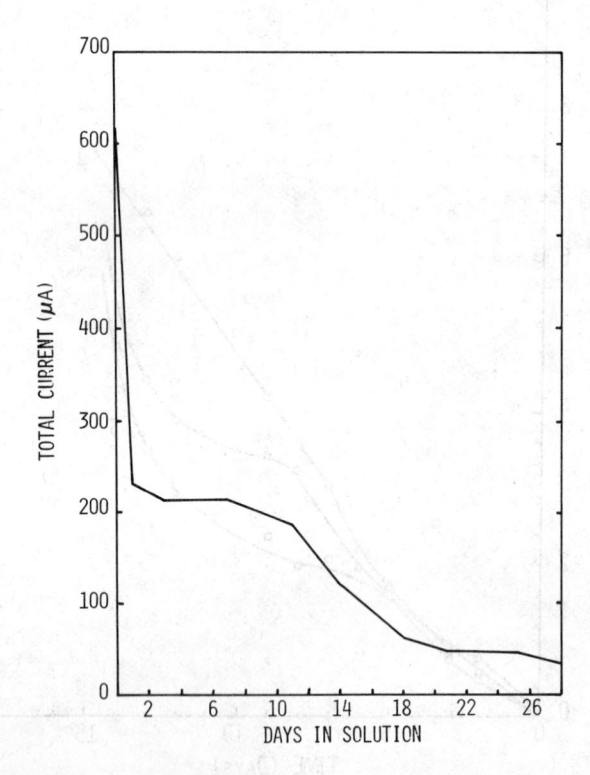

FIG. 4—*Plot of residual current versus time of storage for PHPB films on a 3-mm diameter iron disk electrode reacted with DNPH prior to testing. Deposition and CVs (first sweep) were run in 0.1 M PHPB/0.2 M NaOMe/MeOH.*

125°C (257°F). After two months at temperature, the ampoules were opened, and the specimens were examined. Both the control specimens and unfilmed areas of the TMP/aluminum-zinc specimens were highly pitted. However, those areas which retained the TMP film were remarkably free of pits.

Conclusions

On the basis of our work to date, the following conclusions may be drawn:

1. The adsorption of certain polymer films, electrochemically generated on a metal surface at anodic potentials, protects that surface from corrosion.
2. The amount of protection afforded is largely dependent on the monomer used.
3. Mild steel and copper are readily coated by anodic techniques, and aluminum, despite its oxide layer, is at least partially coated if it is pretreated in a zinc immersion dip.
4. The polymer films deposited have proven to be nonadherent in solutions at elevated temperatures because of the presence of defects or permeability, which allow the solvent to undermine the film and displace it from the metal substrate.
5. Chemical modification of the films may reduce these effects and improve the film stability. Further studies are directed along these lines.

Acknowledgment

This work has been supported by the Solar Heating and Cooling Research and Development Branch, Office of Conservation and Solar Applications, U.S. Department of Energy, under Contract EM-78-C-04-4297.

References

[1] Johnson, S. M. and Simon, F. S., "Comparison of Flat-Plate Collection Performance Obtained Under Controlled Conditions in a Solar Simulator," *Proceedings*, International Solar Energy Society Conference, Vol. 2, Winnipeg, Can., Aug. 1976.
[2] Bier, M., *Electrophoresis*, Academic Press, New York, 1959.
[3] Teng, F. S., Mahalingam, R., Subramanian, R. V., and Raff, R. A. V., *Journal of the Electrochemical Society*, Vol. 124, No. 7, July 1977, pp. 995-1007.
[4] Bruno, F., Pham, M. C., and Dubois, J. E., *Electrochimica Acta*, Vol. 22, 1977, pp. 451-457.
[5] Pham, M. C., Lacaze, P. C., and Dubois, J. E., *Journal of Electroanalytical Chemistry*, Vol. 86, 1978, pp. 147-157.
[6] Maass, W. B., *Metal Finishing*, Vol. 69, 1971, pp. 67-68.
[7] Mengoli, G. and Tidswell, B. M., *Polymer*, Vol. 16, 1975, pp. 881-888.

[8] Keller, F. and Zelley, W. G., *Proceedings of the American Electroplating Society*, Vol. 36, 1949, p. 6.

[9] Garg, B. K., Raff, R. A. V., and Subramanian, R. V., *Journal of Applied Polymer Science*, Vol. 22, 1978, pp. 65-87.

[10] Shreir, L. L., Ed., *Corrosion*, Vol. 1, 2d ed., Newnes-Butterworths, London, 1976, pp. 143-150.

W. J. Dirk,[1] C. A. Allen,[2] and R. E. McAtee[3]

Preliminary Evaluation of Materials for Fluidized Bed Technology in Geothermal Wells at Raft River, Idaho, and East Mesa, California

REFERENCE: Dirk, W. J., Allen, C. A., and McAtee, R. E., **"Preliminary Evaluation of Materials for Fluidized Bed Technology in Geothermal Wells at Raft River, Idaho, and East Mesa, California,"** *Geothermal Scaling and Corrosion, ASTM STP 717,* L. A. Casper and T. R. Pinchback, Eds., American Society for Testing and Materials, 1980, pp. 69–80.

ABSTRACT: Corrosion tests of candidate materials for heat exchangers using liquid fluidized bed technology were conducted at the U.S. Department of Energy's geothermal wells at Raft River, Idaho, and at the Geothermal Components Test Facility (GCTF) in East Mesa, Calif. These investigations deal with moderate temperature ranges of 135 to 163°C for electrical generation and nonelectrical applications. The corrosion, erosion, and scale buildup were evaluated. The average penetration of Type 1025 carbon steel was 314 μm/year. Hastelloy alloys G and C276, stainless steels Type 304L, Type 316, Type 347, and Nitronic 50 had average penetrations of less than 2 μm/year.

KEY WORDS: corrosion, erosion, scaling, materials, evaluation, geothermal

Geothermal water has long been noted for its corrosiveness to a wide range of materials. Geothermal waters vary widely in pH, solid contents, aggressive chemical species, and types of scale formed. Heat exchanger fouling is a major barrier to the development of geothermal systems for electrical energy production. Heat exchanger tubes subject to fouling require at least twice the normal amount of heat transfer surface when compared with heat exchangers [1][4] with no fouling.

Previous studies by Hogg and Grimmett [2] show that a liquid fluidized

[1] Physical chemist, Exxon Nuclear Idaho Co., Idaho Falls, Idaho, 83401.
[2] Branch Manager, Biological and Earth Sciences Branch, E G & G Idaho, Idaho Falls, Idaho 83415.
[3] Senior chemist with E G & G Idaho, Idaho Falls, Idaho 83415.
[4] The italic numbers in brackets refer to the list of references appended to this paper.

bed crystallizer prevents crystals of aluminum nitrate from depositing on cold heat transfer surfaces. The studies also show that heat transfer coefficients between the fluidized bed and the immersed heat transfer surfaces are higher than those obtained without a fluidized bed. The movement of the particles past the surfaces reduced the film thickness and allowed the increased rate of heat exchange. These two observations led to the suggestion that fluidized bed heat exchangers could be developed that would prevent the usual scaling caused by geothermal brines cooled by contact with heat exchanger surfaces. Because of the increased heat transfer rate observed, less heat transfer surface would be required for a given capacity than that needed in a conventional exchanger.

The experimental heat exchanger design developed for testing utilizes the geothermal brine as the fluidizing liquid on the shell side of the exchanger, while the secondary fluid passes through the heat exchanger tubes (Fig. 1).

Experimental Procedure

Selection of the materials for the experimental heat exchanger construction was based on materials that were commercially available and cost effective. The materials chosen were Type 304L, Type 316, Type 347, and Nitronic 50 stainless steels, Hastelloy alloys G and C-276, and Type 1025 carbon steel. Nominal compositions of the materials tested are listed in Table 1. The materials chosen exhibited characteristics of corrosion, erosion, and scaling resistance to a wide range of solutions [3,4,5].

The experimental heat exchanger used in the experiment was a tube and shell design. Since the fluidized bed is isothermal, the unit could be considered one stage in a multistage heat exchanger. It was equipped with view glasses so that observations could be made of the vessel interior during the experiment. Temperature probe measurements, not included in this paper, were taken and monitored continuously so that bed gradients and heat balances could be determined.

Test coupons, 0.64 by 2.54 by 5.08 cm, were machined to a 125 root mean square (rms) surface finish from strips welded by the manual gas tungsten arc welding (GTAW) process. The welded coupons were picked to evaluate corrosion due to possible sensitization of the heat-affected zone (HAZ) in the austenitic stainless steels. These coupons were electrically isolated and separated by the use of Pyrex tubing, and spacers and were suspended in the experimental heat transfer system to allow vertical flow of the geothermal well water. The lower set of coupons was suspended vertically in the fluidized bed; the upper set was suspended vertically in the disengaging space above the fluidized bed. The coupons were suspended to allow a leading edge and a vertical surface to be exposed to the fluidized bed. A stainless steel wire insulated with Teflon heat-shrinkable tubing was

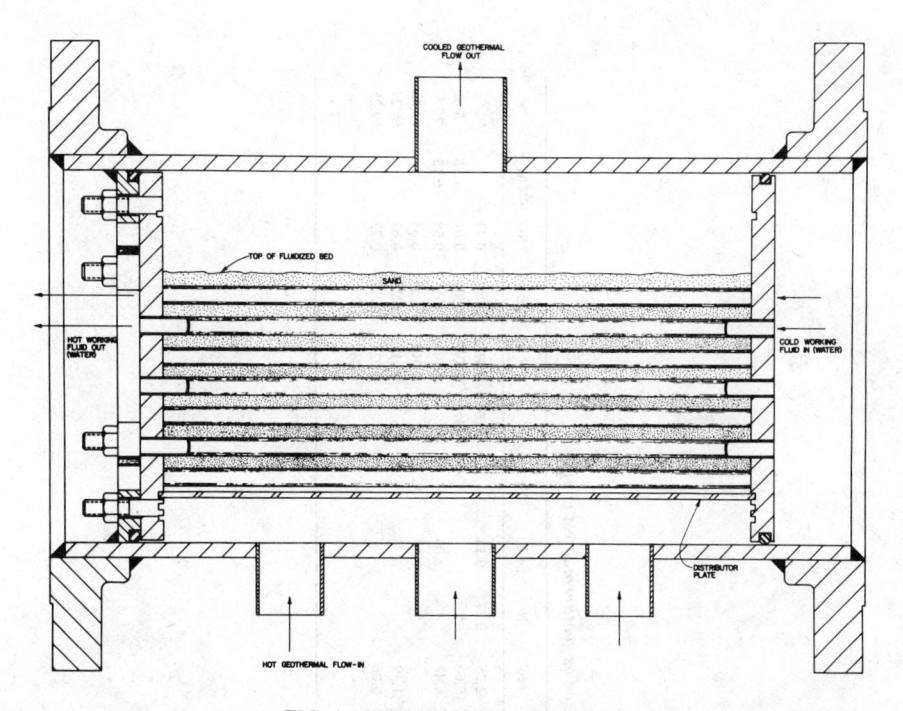

FIG. 1—*Fluidized bed heat exchanger.*

used to attach the coupons to the heat exchanger. The coupons were inspected at ×60 magnification prior to and after the testing (Fig. 2).

Test at Raft River Geothermal Well No. 1

The water composition, shown in Table 2, was determined for Raft River geothermal well RRGE-1.

In the first test, the heat exchanger was run for a period of 62 days. The coupons of Type 304L, Type 347, and Nitronic 50 stainless steels and of Hastelloy alloy C-276 exhibited a light, clear, brown deposit. This deposit was removed ultrasonically, filtered, and analyzed by X-ray diffraction and emission spectroscopy. The crystalline component of the specimen was identified as muscovite [$KAl_2(Si_3Al) O_{10} (OH, F_2)$]. Muscovite is a member of the mica group occurring in white, green, red, or light-brown forms. Muscovite was also found in one of the liquid samples collected under pressure during this test. The corrosion rates of the test samples were <0.3 μm/year in the disengaging space. The large differential between the fluidized bed and the disengaging space is probably due to erosion by bed particles. During operation, high-velocity channeling of the bed material

TABLE 1—Composition of materials used as geothermal test coupons.

Material	C	Mn	Cu	Cr	Co	Ni	Mo	N	Cb	Ta	V	P	S	Si	W	Fe
Nitronic 50 stainless steel	0.05	5.30		21.10		13.50	2.25	0.27	0.14		0.18	0.027	0.012	0.47		5.86
Hastelloy alloy C-276	0.004	0.48		16.0	2.24	55.69	15.60				0.31	0.02	0.006	0.01	3.78	20.5
Hastelloy alloy G	0.028	1.52		22.89	0.71	45.07	6.47		2.09			0.007	0.005	0.58	0.13	67.65
Type 347 stainless steel	0.057	1.42		18.45		10.65			0.57	0.02		0.013	0.019	0.64		66.24
Type 316 stainless steel	0.020	1.68	0.08	17.24		12.87	1.19					0.018	0.029	0.63		69.25
Type 304L stainless steel	0.028	1.53	0.06	18.28	0.20	9.83	0.06					0.027	0.016	0.72		99.03 to
Type 1025 carbon steel	0.21 to 0.28	0.3 to 0.6										0.04	0.05			99.4

FIG. 2—*Raft River geothermal well No. 1 test coupons.*

TABLE 2—*Major chemical species found in Raft River well RRGE-1.*

Ca^{++}	53.5[a]
F^-	6.3
Fe	0.1
HCO_3^-	63.9
$SO_4^=$	60.2
Cl^-	776
O_2	0.00013
Total dissolved solids	1560.0
pH	6.27

[a] Figures represent mg/litre, except for pH.

caused erosion of the carbon steel heat exchanger shell (Fig. 3). This problem was corrected by a change in the distributor plate design in a later heat exchanger.

A second set of test coupons was installed in a heat exchanger in which the distributor plate was modified to prevent bed channeling. The list of materials tested included those from the first test plus Type 316 stainless

FIG. 3—*Erosion of carbon steel fluidized bed shell.*

steel, Hastelloy alloy G, and unwelded Type 1025 carbon steel. The test coupons were run for 984 h and then examined. A very light coating of what appeared to be muscovite was noted on the coupons in the disengaging space. An insufficient amount of this substance was available for analysis.

The test coupons were examined for specific types of attack. Pitting attack was <25 μm in depth. The heat exchanger tubes constructed of Type 304L stainless steel exhibited pitting. Further testing is recommended to confirm whether stress corrosion cracking (SCC) is a problem in this environment. The Type 1025 carbon steel coupons exhibited a black tenacious coating, which was analyzed by X-ray diffraction and identified as iron oxide and magnetite (FeO, Fe_2O_3, Fe_3O_4). The corrosion rates and types of attack for the fluidized bed and the disengaging space above the bed are listed in Table 3. Figure 4 is a photograph of the test coupons after an exposure of 984 h.

Tests at the Geothermal Components Test Facility, East Mesa, Calif.

The first test performed at the Geothermal Components Test Facility (GCTF) at East Mesa, Calif., was done with glass bench test equipment.

TABLE 3—*Corrosion rates of test coupons exposed to Raft River well RRGE-1 for 984 h.*

Alloy	Type of Corrosion[a] Above Fluidized Bed	Type of Corrosion[a] In Fluidized Bed	Corrosion Rate, μm/year Above Fluidized Bed	Corrosion Rate, μm/year In Fluidized Bed
Type 304L stainless steel	2	2	0.81	1.94
Type 316 stainless steel	2	2	0.54	0.76
Type 347 stainless steel	2	2	0.84	0.91
Hastelloy alloy C-276	1	1	0.56	0.70
Hastelloy alloy G	1	1	0.37	0.13
Nitronic 50 stainless steel	1	1	0.42	0.56
1025 carbon steel	5,6	2	20.13	44.62

[a] Type of corrosion: (1) no visible corrosion, (2) light pitting, (3) light etch, (4) light crevice attack, (5) moderate pitting, (6) end grain attack, (7) surface very irregular, (8) general attack.

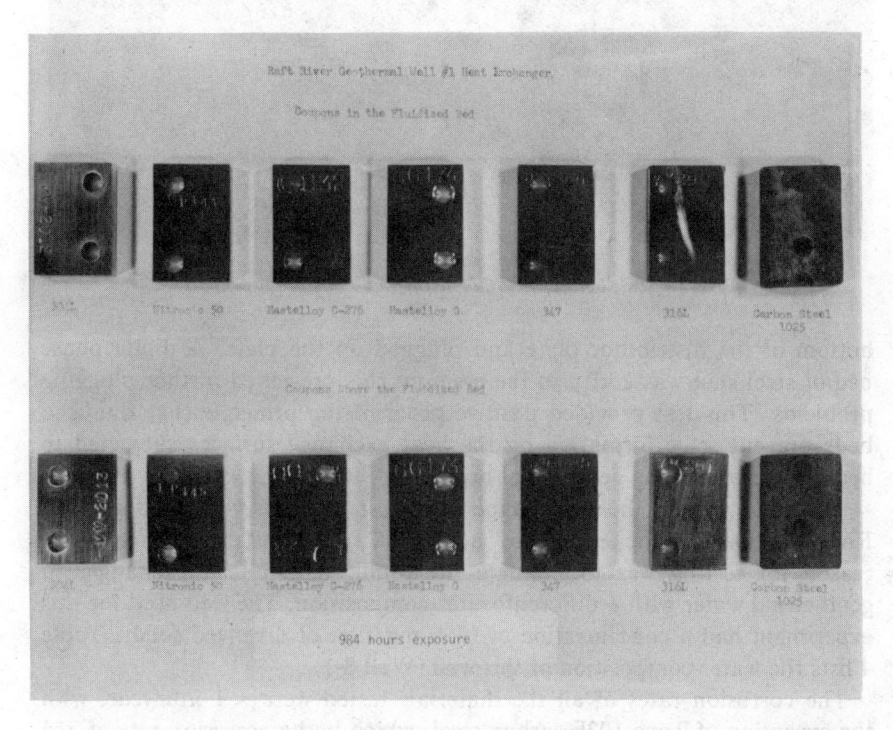

FIG. 4—*Test coupons from Raft River geothermal well No. 1 after 984 h of exposure.*

No corrosion coupons were used in water from Well 6-1, which had a dissolved solids content of 2500 mg/litre. Calcite scale formed in the lines from the flash tank to the heat exchanger. A black scale, identified as a combination of iron oxide and magnetite, formed in the lines from the flash tank to the heat exchangers (Fig. 5). Black scale formed on the

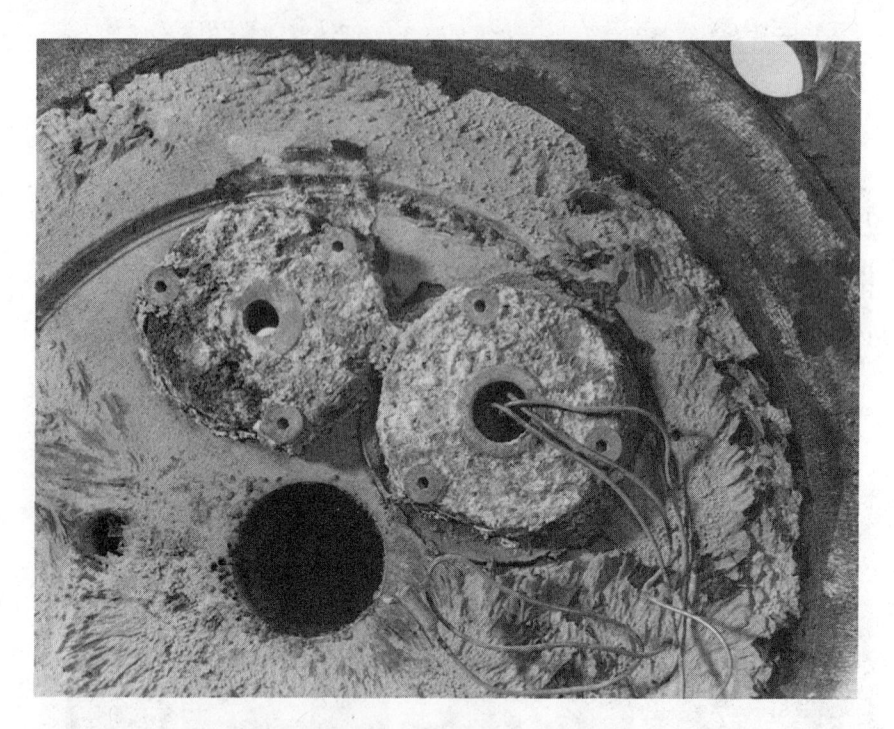

FIG. 5—*Scale buildup at GCTF in the water inlet from Well 8-1.*

bottom of the distributor plate and plugged up the plate. A dilute phase bed of steel shot was added to the plenum; this prevented further plugging problems. This test provided positive proof of the principle that fluidized beds prevent scale formation on the heat exchange surfaces subjected to heavy scale-producing geothermal brine.

The same type of heat exchanger and test setup as that used at Raft River was used in the second test at the GCTF in East Mesa, Calif. The reason for testing was to evaluate the fluidized bed heat exchanger in geothermal water with a different water composition. The well used for this experiment had a concentration of 1750 mg/litre of dissolved solids. Table 4 lists the water composition of this well (Well 8-1).

The corrosion rates of all the materials tested were <1 μm/year, with the exception of Type 1025 carbon steel, which had a corrosion rate of 314 μm/year. The corrosion rates and types of attack are listed in Table 5. Figures 6 and 7 are photographs of the test coupons after 2184 h of exposure.

Results and Discussion of Corrosion Tests

The materials tested at Raft River and East Mesa are suitable for use as heat exchanger materials, with the exception of Type 1025 carbon steel.

TABLE 4—*Major chemical species found in the GCTF well, East Mesa, Calif.*

Ca^{++}	8.5[a]
F^-	1.6
Fe	<0.1
HCO_3^-	417.0
$SO_4^=$	173
Cl^-	500
Aluminum	0.02
Total dissolved solids	1750
pH	6.27

[a] Figures represent mg/litre, except for pH.

TABLE 5—*Corrosion rates of test coupons exposed to the GCTF well East Mesa, Calif. for 2184 h.*

	Type of Corrosion[a]		Corrosion Rate, μm/year	
Alloy	Above Fluidized Bed	In Fluidized Bed	Above Fluidized Bed	In Fluidized Bed
Type 304L stainless steel	1,2	1	0.72	0.24
Type 316 stainless steel	1,2	1,2	0.48	0.60
Type 347 stainless steel	1,2	1,2	0.96	0.36
Hastelloy alloy C-276	1,2	1,2	0.48	0.48
Hastelloy alloy G	1	1,2	0.36	0.12
Nitronic 50 stainless steel	1,2	1,2	0.36	0.48
1025 carbon steel	6	3,4,5	250.8	314.4

[a] Type of corrosion: (1) light etch, (2) light crevice attack, (3) end grain attack, (4) pitting attack, (5) surface very irregular, (6) general attack.

The pitting and crevice corrosion noted were in all cases <25 μm in depth.

Chemical analysis of the water samples exhibited low Cl^- and oxygen concentrations. Typically, low-chloride deaerated solutions are less aggressive than those with moderate amounts of oxygen (>0.5 ppm). Chlorides in the presence of oxygen can cause severe pitting of austenitic stainless steels and, in some cases, can promote stress corrosion cracking (SCC) at elevated temperatures [6,7]. Application of these materials to other geothermal water sources should be verified by a test program.

Conclusion

Heat exchanger materials were evaluated on the basis of weight loss, resistance to pitting, and crevice corrosion. They were rated, in the order of most to least desirable, as follows: Hastelloy alloy G, Nitronic 50 stainless steel, Hastelloy alloy C-276, stainless steel Types 347, 316, and 304L,

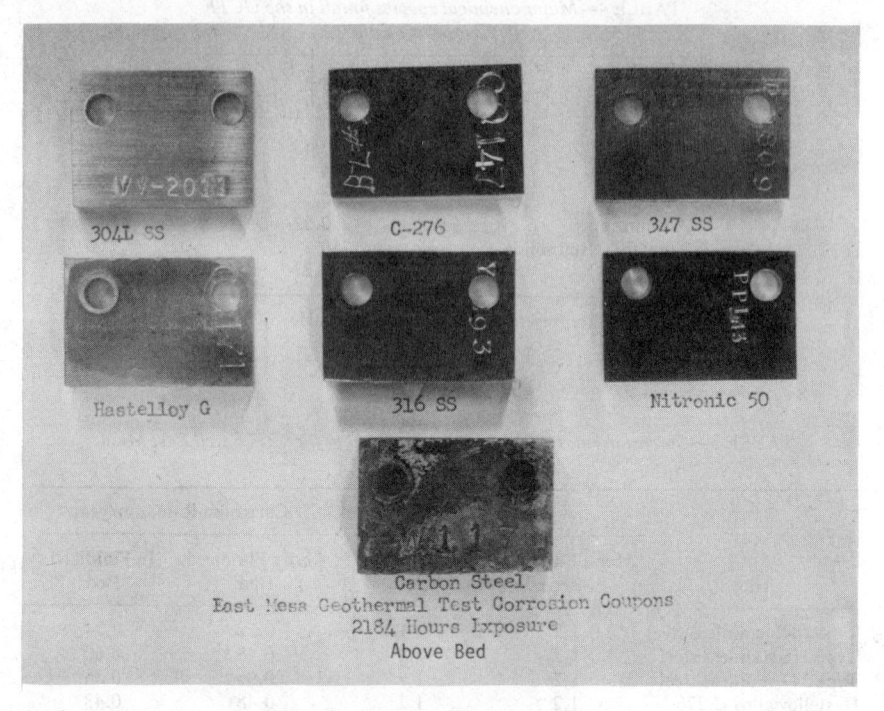

FIG. 6—*Corrosion coupons from the East Mesa geothermal test after 2184 h of exposure above the bed.*

and carbon steel Type 1025. Type 1025 carbon steel was the only material found unsuitable for use as a heat exchange material.

Scaling problems of the heat exchanger tubes were essentially eliminated with the use of the fluidized bed design in this test.

Summary

This report describes the results and method of corrosion testing of candidate materials for use in heat exchangers using liquid fluidized bed technology. The testing was conducted at the U.S. Department of Energy's (DOE) geothermal wells at Raft River, Idaho, and at the Geothermal Components Test Facility (GCTF) in East Mesa, Calif.

The purpose of these tests was to evaluate erosion, corrosion, and scaling of the fluidized bed heat exchanger. Two sets of coupons were run in these tests. One set was located in the fluidized bed to evaluate erosion corrosion. The second set of coupons was located in the disengaging space above the fluidized bed to evaluate corrosion and scaling potentials.

A light-brown scale was removed from the Raft River test coupons

FIG. 7—*Coupons from the East Mesa geothermal test after 2184 hours of exposure in the bed.*

located in the disengaging space above the fluidized bed. This material was analyzed by X-ray diffraction and emission spectroscopy and identified as muscovite [$KAl_2(Si_3Al)O_{10}(OH,F_2)$]. A material with the same general appearance was noted for the coupons in the GCTF well, but in insufficient amounts for analysis. The test coupons in the fluidized bed of the Raft River and GCTF wells had a clean, polished appearance, with the exception of small amounts of scale at the bottom of the stamped coupon numbers.

The welded test coupons were evaluated after 984 h of exposure in Raft River geothermal well RRGE-1 and after 2184 h of exposure at the GCTF in East Mesa, Calif. Type 304 L, Type 316, Type 347, and Nitronic 50 stainless steels and Hastelloy alloys G and C-276 had corrosion rates of <2 μm/year. Type 1025 carbon steel exhibited a corrosion rate of 314 μm/year.

Acknowledgment

Funding for the development and application of a liquid fluidized bed heat exchanger for geothermal applications was provided by the U.S.

Atomic Energy Commission, Energy Research and Development Administration, and the U.S. Department of Energy, in succession.

References

[1] Allen, C. A., Fukuda, O., Grimmett, E. S., and McAtee, R. E., "Liquid Fluidized Bed Heat Exchanger—Horizontal Configuration—Experiment and Data Correlation," 12th Intersociety Energy Conversion Engineering Conference, Washington D.C., 28 Aug. 1977, pp. 824-831.
[2] Hogg, G. W. and Grimmett, E. S., *Chemical Engineering Progress Symposium Series*, Vol. 62, No. 67, 1966, pp. 51-56.
[3] Manrer, J. R., "New Austenitic and Ferritic Stainless Steels for Geothermal Applications," Geothermal Resource Council, Davis, Calif., workshop on Material Problems Associated with the Development of Geothermal Energy Systems, El Central, Calif., 16-18 May 1975, pp. 107-132.
[4] Committee of Stainless Steel Producers, "The Role of Stainless Steels in Desalination," Publication No. 4745 M, April 1974, pp. 1-32.
[5] Tuthill, A. H. and Schillmoller, C. M., International Nickel Co., "Guidelines for Selection of Marine Materials," Paper presented at the Ocean Science and Ocean Engineering Conference, Marine Technology Society, Washington, D.C., 14-17 June 1965.
[6] Szklarska-Smialowska, S., *Localized Corrosion NACE 3*, Dec. 1971, p. 312.
[7] Moller, G. E., *Society of Petroleum Engineers*, Vol. 17, No. 2, 1977, pp. 101-110.

W. R. Braithwaite[1] and K. A. Lichti[2]

Surface Corrosion of Metals in Geothermal Fluids at Broadlands, New Zealand

REFERENCE: Braithwaite, W. R. and Lichti, K. A., "**Surface Corrosion of Metals in Geothermal Fluids at Broadlands, New Zealand,**" *Geothermal Scaling and Corrosion, ASTM STP 717*, L. A. Casper and T. R. Pinchback, Eds., American Society for Testing and Materials, 1980, pp. 81-112.

ABSTRACT: Corrosion studies of metallic construction materials for a geothermal power plant at Broadlands, New Zealand, are described. Surface corrosion investigations on 36 materials in seven geothermal media, using the ASTM coupon method, have been initiated, and results for selected materials are presented in a logarithmic graphical form. The observed corrosion characteristics are discussed in terms of corrosion product composition and stability and theoretical film-formation mechanisms.

KEY WORDS: corrosion, reaction kinetics, corrosion mechanisms, iron sulfides, carbon steels, scaling, alloy steels, titanium alloys, copper alloys, cleaning, geothermal

Corrosion investigations of materials proposed for use in a geothermal power plant in the Broadlands, New Zealand, hydrothermal area began in 1974. Materials previously tested at the Wairakei geothermal field in New Zealand [1][3] had to be reevaluated at the Broadlands field because of the higher gas content [2]. In addition, a study of improved turbine blading materials, superior to the traditional 13 percent chromium, low-carbon stainless steels, was considered essential. Other materials for special applications in geothermal environments also required field testing before use.

The present corrosion studies at Broadlands are based on insulated coupon weight losses and simple stress corrosion tests, using restricted U-bend specimens exposed in static environments. Other tests are being conducted

---
[1] Scientist, Chemistry Division, Department of Scientific and Industrial Research, Petone, New Zealand. Braithwaite is the investigator who initiated the project.
[2] Scientist, Industrial Processing Division, Department of Scientific and Industrial Research, Petone, New Zealand. Lichti is the investigator responsible for preparing this paper and the author to whom all correspondence should be addressed.
[3] The italic numbers in brackets refer to the list of references appended to this paper.

to study stress corrosion and hydrogen-induced cracking, using notched stress-rupture specimens and corrosion fatigue, using Wöhler-type specimens.

This paper provides a progress report on current results from coupon weight-loss surface corrosion tests.

Description and Operation of Corrosion Rig

Surface corrosion investigations were conducted in seven environments typically encountered in geothermal fluid transmission systems, steam turbines, and condensing systems. A corrosion rig was constructed on site for static exposures of metal coupons in bore fluid, separated bore water, high-pressure steam, wet low-pressure steam, wet low-pressure aerated steam, cold condensate, and aerated cold condensate. Figure 1 gives a schematic representation of the corrosion rig, and Table 1 describes the physical conditions within the test vessels.

Bore fluid was drawn from Well BR22 (Fig. 1), and gate-type control valves were used to sample some of this fluid for corrosion testing and to direct the remainder to a cyclone separator. Separated bore water was also sampled for corrosion testing, and the separated high-pressure steam was fed to a manifold, which supplied dry steam for the rest of the plant. Stable operating conditions for the corrosion tests were governed by manual pressure control. The bore-head pressure of 3.03 MPa was reduced to 1.39 MPa for transmission and separation of the bore fluid. The separated steam manifold was operated at 650 kPa. Vessels containing bore fluid (*1*), separated bore water (*2*), and separated high-pressure steam (*3*) were all maintained at 650 kPa. Conditions in the wet low-pressure steam vessels (*4* and *5*) were obtained by mixing controlled amounts of wet and dry steam. Wet steam was obtained with the aid of a boiling water-type condenser operated at the boiling point. The wetness fraction was monitored daily, using the calorimetric method outlined by Lyle [*3*]. A flowmeter and control valve were used to introduce 3 percent by volume of air to the wet steam of Vessel 5 for accelerated steam corrosion tests. Water-cooled coil condensers were used to condense and cool separated steam for the liquid condensate vessels, Vessels 6 and 7. Three percent by volume of air was added to the steam for Vessel 7 before condensation to provide an aerated condensate corrosion environment.

Vessels 1 to 5 were constructed from 254.5-mm (10-in.) American Petroleum Institute high-test line pipe. Mild steel flanges and bolted face plates with asbestos gaskets gave access to the vessels that were supported horizontally. The condensate pressure vessels (6 and 7) were vertical to prevent fluid loss on removal of the lids. Prepared coupons were loaded onto racks, as seen in Fig. 2, which were located on supporting rails in the horizontal vessels and on specially designed holders in the vertical vessels. All of the

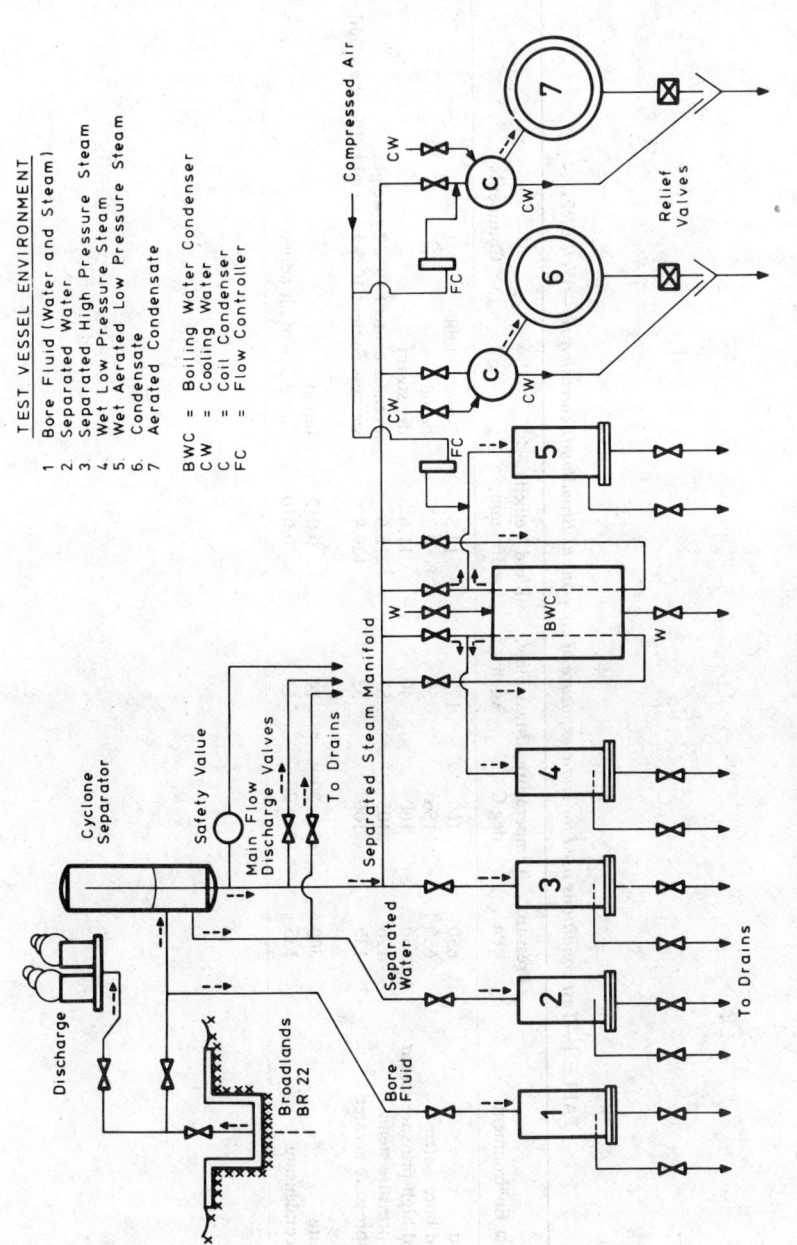

FIG. 1—Schematic representation of Broadlands, New Zealand, corrosion rig.

TABLE 1—*Test conditions used for static exposure of coupons at Broadlands corrosion rig—Well BR22.*

Test Environment	Pressure, kPa	Temperature, deg C	Mass Flow, kg/h	Fluid Velocity, dm/min	Comments
Bore fluid	650	160	178	14.5	2-phase fluid
Separated bore water	650	156	56	0.2	liquid
Separated high-pressure steam	650	160	40	11.6	dry steam
Wet low-pressure steam	126	105	57	215.6	wetness content 10% by weight
Wet low-pressure aerated steam	135	106	32	129.9	wetness content 10% by weight, 3% of air added
Condensate	300	10 to 20	14	0.012	liquid
Aerated condensate	135	10 to 20	12	0.010	liquid 3% of air added

vessels had the capacity for three racks with the dimensions 229 by 305 mm. Diffuser plates were used to prevent direct impingement of incoming fluids on exposed coupons. The racks were inserted and removed with the vessels at operating temperature. After exposure, the racks holding the coupons were washed at once in hot water and dried at 110°C. After cooling, the racks were stored in polyethylene bags containing a desiccant, until disassembled under controlled laboratory conditions.

Test Environment Chemistry

The chemistry of the Broadlands geothermal area has been reported by Mahon and Finlayson [2,4]. The bore from which steam was drawn for the corrosion tests (Well BR22) had total discharge gas concentrations of 234 mmol/100 mol of carbon dioxide (CO_2) and 5.14 mmol/100 mol of hydrogen sulfide (H_2S) with a CO_2/H_2S ratio of 45.5 and a total heat capacity of 1230 J/g at 1.3 MPa. Separated steam at 690 kPa had 2.3 percent by weight of gas ($CO_2 + H_2S$) [4]. The available chemical data for the corrosion test environments are summarized in Table 2 [5]. A detailed report on chemical analysis of the geothermal fluids tested at Broadlands, including a critical review of fluid sampling and analysis techniques, is in preparation [5].

Experimental Procedures

Corrosion tests were conducted on 36 ferrous and nonferrous alloys (Table 3) in a program that included three exposure periods (4 weeks, 13 weeks, and 52 weeks). Some preliminary tests, up to 13 weeks, were made on Type 1010 carbon steel and Types 304 and 410 stainless steels (Table 3) to check the corrosion rig control systems and to establish the order of weight losses to be expected.

Test coupons having a minimum exposed surface area of 0.23 dm^2 (typically 0.55 dm^2) were obtained and prepared according to the ASTM Recommended Practice for Preparing, Cleaning, and Evaluating Corrosion Test Specimens (G 1-72) and the ASTM Recommended Practice for Conducting Plant Corrosion Tests [G 4-68 (1974)]. After degreasing in acetone and trichloroethylene vapor, the coupons were coded, using 3-mm-high letters and numbers. Abrasive cleaning methods were used to remove all previously formed oxides or films. The stainless steel alloys were given a surface conditioning using wet 240-grit silicon carbide paper to avoid smooth-surface effects, whereas all other materials were scrubbed under a slurry of 300-mesh pumice with a bristle brush. The lead coupons were mechanically scraped and then polished with a soft cloth. The treatments were completed by hot and cold distilled water washing, acetone washing, and warm-air drying, followed by swabbing and rinsing in hexane to remove residual cleaning materials, and then followed by 100°C air drying. The coupons were then

TABLE 2—*Summary of corrosion test environment chemical conditions* [5].

Test Environment	pH	Condensed Fluid Conductivity, S/m × 10^{-2}	Chemical Species, mmol/litre				
			H_2S	SO_4	NH_3	CO_2	Fe
Separated water	7.5	40.0	0.32	0.10	0.28	?	nil
HP steam[a]	5.7	2.50	9.5	nil	2.35	475	nil
Wet low-pressure steam			9.6[b]				
Wet low-pressure aerated steam			11.4[b]				
Condensate	5.15	1.92	2.4	0.004	2.35	48.2[c]	0.226
Aerated condensate	6.0	2.33	0.26	0.14	2.24	8.1[c]	0.204

[a] Sampled before passing through the corrosion vessel.
[b] High H_2S in aerated steam due to the collection method, during which steam was lost.
[c] Calculated.

TABLE 3a—*Composition of metals*

Metal or Alloy	Equivalent Specification, UNS[a]	Specimen Exposures[b]	Ferrous Metals and Alloys—			
			C	Mn	P	S
Carbon steel	G10180 (*CS*)[d]	T	0.15 to 0.19	0.62	...	...
Low-alloy steel	C41400 (*LAS*)[e]	T	0.41	0.82	...	...
Cast iron	Grey (*CI*)	D	3.3	0.62	...	...
Cast iron	*Ni-Resist* type *1*	D	2.8	1.1	...	...
Stainless steel	S*41000*	T	0.14	0.37	0.012	0.006
Stainless steel	*E-Brite 26-1*	D	<0.01	0.08	...	...
Stainless steel	*FV520 (B)*	T	<0.05	0.80	...	...
Stainless steel	Sandvik *3RE60*	D	0.04	1.5	...	...
Stainless steel	S*30400*	T	0.05	1.45	0.026	0.009
Stainless steel	S*31600*	T	0.067	1.65	0.030	0.017
Stainless steel	S*32100*	D	0.05	1.54	0.028	0.012
Stainless steel	S*34700*	D	0.059	1.58	0.030	0.015
Stainless steel	Sandvik *2RK65*	D	<0.01	1.5	...	...
Stainless steel	N*08020* (Carp *20Cb3*)	T	0.04	0.76	0.022	0.005
Ni/Fe/Cr alloy	N*08825* (Incoloy*825*)	T	0.02	0.33	...	0.003
Preliminary Trials						
Carbon steel	G10100	D	0.08	0.35	...	...
Stainless steel	S41000	D	0.087	0.19	...	...
Stainless steel	S30400	D	0.08	1.34	...	...

[a] The italic portion of the specification is used to identify materials on graphs. UNS = unified numbering system.
[b] Abbreviations used: D = duplicate specimen exposures; T = triplicate specimen exposures.
[c] Analysis for heat exposed.

cooled in circulating dry air and were stored under desiccation prior to repeated weighings. The coupons for the main program were weighed to 0.01-mg accuracy, and for the preliminary tests, to 0.1 mg accuracy. Triplicate specimens were used for important materials and duplicates for all others (see Table 3). The typical center-line average-surface roughness values for the exposed materials were 0.2 μm, for mild steel, low-alloy steel, the cast irons, tantalum and molybdenum; 0.5 μm for stainless steels, nickel alloys, the cobalt alloy, and titanium alloys; 2.0 μm for copper, brass, lead, and zinc; and 4.0 to 5.0 μm for Alloy E-Brite 26-1 and zirconium Alloy GR32.

Racks were made up of prepared coupons, supported on rods sleeved with polytetrafluoroethylene (PTFE) and separated by insulating spacers of the same material (see Fig. 2). The loaded racks were transported to the test site in sealed polyethylene bags containing desiccant.

After the test exposure, the coupons were photographed and individually studied before being weighed. Loose corrosion products were mechanically removed for X-ray analysis, and chemical cleaning methods were applied to remove adherent corrosion products, as outlined in Table 4 [6,7,8,9,10].

and alloys exposed at Broadlands.

Element, Weight Percent[c]

Si	Ni	Cr	Mo	V	Cu	Others
<0.02	0.03	0.04	<0.03	...	...	...
0.28	0.21	0.87	0.19	...	0.22	...
2.6	0.06	0.07	<0.03	0.02	0.06	...
2.8	14.9	2.1	0.03	...	6.5	0.15 Al; 0.13 Co
0.40	0.18	12.36	0.080	...	...	0.005 Al; 0.006 Sn
0.26	0.18	25.1	1.26	0.11	0.04	0.02 Co; 0.01 Ti; <0.02 Zr
0.44	5.1	14.3	1.58	0.05	1.45	0.05 Al; 0.05 Co; <0.02 Ti
1.1	5.1	19.5	3	0.04	0.09	0.1 Co; 0.02 Ti
0.70	9.10	18.50	...	...	...	...
0.46	13.31	16.48	2.17	...	...	...
0.69	9.60	17.59	0.35	...	0.25	0.66 Ti; 0.17 Co
0.67	10.30	18.50	0.17	...	0.16	0.61 Cb + Ta
0.4	24	20	5.5	0.03	1.4	0.2 Co; 0.02 Ti
0.38	31.73	19.57	2.27	...	3.21	0.64 Cb + Ta
0.24	42.96	22.36	2.97	...	1.87	0.08 Al; 0.91 Ti; 28.24 Fe
0.005	0.02	0.04	<0.02	<0.005	0.04	
0.32	0.25	13.6	<0.01	<0.02	0.13	
0.79	8.8	19.0	0.18	0.035	0.06	0.03 Ti; 0.02 Co

[d]Carbon steel G10180 had a grain orientation parallel to the exposed face of the coupon.
[e]Low alloy steel G41400 had a grain orientation perpendicular to the exposed face of the coupon.

TABLE 3b—*Composition of metals*

Metal or Alloy (D)[a]	Equivalent Specification, UNS[b]	C	Mn	Si	Fe
Aluminum alloy	A9*1100*	...	0.05 max	1.0 Si + Fe max	
Al/Mg alloy	A9*5052*	...	0.10 max	0.45 Si + Fe max	
Al/Mg/Si alloy	A9*6061*	...	0.15 max	0.4 to 0.3	0.7 max
Al/Mg/Si alloy	A9*6063*	...	...	0.2 to 0.6	0.35 max
De-Oxidised Copper[c]	C*10200*	...	...	...	...
Aluminum Brass (As)[c]	C*68700*	...	0.022	0.02	0.01
Nickel 200	N0*2200*	0.15 max	0.35 max	0.35 max	0.40 max
Ni/Cu-Monel	N0*4400*	0.3 max	2.00 max	0.50 max	2.5 max
Ni/Cu-Monel K500[c]	N0*5500*	0.18	0.64	0.12	...
Ni/Cr-Inconel 600[c]	N0*6600*	0.02	0.20	0.19	6.25
Ni/Mo-Hastelloy B[c]	N*10001*	0.014	0.49	0.29	5.19
Ni/Mo-Hastelloy C276[c]	N*10276*	0.002	0.47	0.03	6.32
Cobalt Alloy	Stellite *6B*	1	...	...	...
Titanium[c]	*Ti*	0.026	...	...	0.06
Ti-6Al-4V[c]	IMI *318*	...	...	...	0.10
Ti-4Al-4Mo-2Sn-0.5 Si[c]	IMI *550*	...	...	...	...
Molybdenum	*Mo*	0.01 to 0.04	...	0.10 max	0.10 max
Tantalum[c]	*Ta*	...	...	...	...
Zirconium alloy[c]	GR32 (*Zr*)	...	...	...	0.13
Lead	*Pb*	...	...	...	...
Zinc	*Zn*	...	...	...	0.02 max

[a](D) = duplicate specimen exposures of all alloys shown.
[b]The italic portion of the specification is used to identify materials on graphs. UNS = unified numbering system.

After successive chemical cleanings, the loosened corrosion products were scrubbed off with a wet nylon bristle brush. The cleaned coupons were washed in cold and hot distilled water, dried in warm air, and then vacuum-desiccated before being weighed. The effectiveness of the cleaning was judged visually and supported by weight-loss measurements. Weight-loss results were obtained for one of two or two of three coupons exposed in each environment prior to deciding whether the remaining coupon should be used for metallographic examination or additional weight-loss evaluation.

Results of Preliminary Trials

Figure 3 shows the complete set of coupons exposed in the preliminary tests. Adherent uniform films of corrosion product were formed on mild steel coupons in bore fluid, separated bore water, and nonaerated condensate. This was in contrast to the bulky scales formed on mild steel in high-pressure steam and wet low-pressure steam. A thick massive corrosion product, which

and alloys exposed at Broadlands.

Metals—Element, Weight Percent					
Ni	Cr	Al	Ti	Cu	Others
...	...	99.00	...	0.05 to 0.20	0.10 max Zn
...	0.15 to 0.35	REM[d]	0.15 max	0.10 max	2.2 to 2.8 Mg; 0.10 max Zn
...	0.40 to 0.35	REM	0.15 max	0.15 to 0.40	0.8 to 1.2 Mg; 0.25 max Zn
...	...	REM	0.10 max	0.10 max	0.45 to 0.9 Mg; 0.10 max Zn
...	...	...	...	99.95 min	(including Ag)
<0.01	...	2.1	...	REM	0.05 As; 0.005 Sn; 20.5 Zn
99.0 max	...	...	...	0.25 max	0.010 max S
63.00 to 70.00	...	...	...	REM	0.024 max S
65.34	...	2.81	0.50	29.42	0.005 S
77.67	15.60	...	...	0.04	0.007 S
REM	0.63	...	...	...	27.06 Mo; 0.011 P; 0.001 S; 0.41 Co; 0.35 V
REM	15.07	...	...	...	6.32 Mo; 0.024 P; 0.012 S; 1.94 Co; 3.09 W
...	26	...	...	...	68 Co; 5 W
...	...	...	REM	...	0.01 N; 0.13 O; 0.006 H
...	...	6.1	REM	...	3.9 V
...	...	3.7	REM	...	3.7 Mo; 1.8 Sn; 0.39 Si
0.005 max	...	...	...	...	99.9 min Mo
...	...	...	...	...	99.86 min Ta
0.05	0.08	...	...	...	1.57 Sn; 98.16 Zr
...	...	...	...	Cu + Ag 0.0025 max	99.94 Pb; 0.001 max Zn; 0.002 max As + Sb + Sn
...	...	...	...	0.005 max	REM Zn; 0.12 to 0.20 Cd; 0.20 to 0.50 Pb

[c] Analysis for heat exposed.
[d] REM = remainder of composition (or balance).

was blistered and porous formed in aerated wet steam. In aerated condensate a thick deposit of nonadherent corrosion product formed, which easily washed off to reveal bright metal. The corrosion products observed on the stainless steels (Fig. 3) were generally thin and adherent. Some nonuniform staining or tarnishing, however, was evident on coupons exposed in bore fluid and the steam environments. This was most pronounced on coupons exposed to aerated wet steam. Stainless steel coupons exposed in aerated condensate appeared to have deposits of foreign corrosion products in addition to the surface films.

After being cleaned and weighed, the coupons were inspected under oblique light with ×10 binocular magnification. The surfaces of the mild steel coupons from all environments were generally smooth with some pitting. The coupons exposed in wet low-pressure aerated steam and non-aerated condensate exhibited large-diameter deep pits with depths of 1 and 0.5 mm, respectively. In aerated condensate, corrosion was very even with

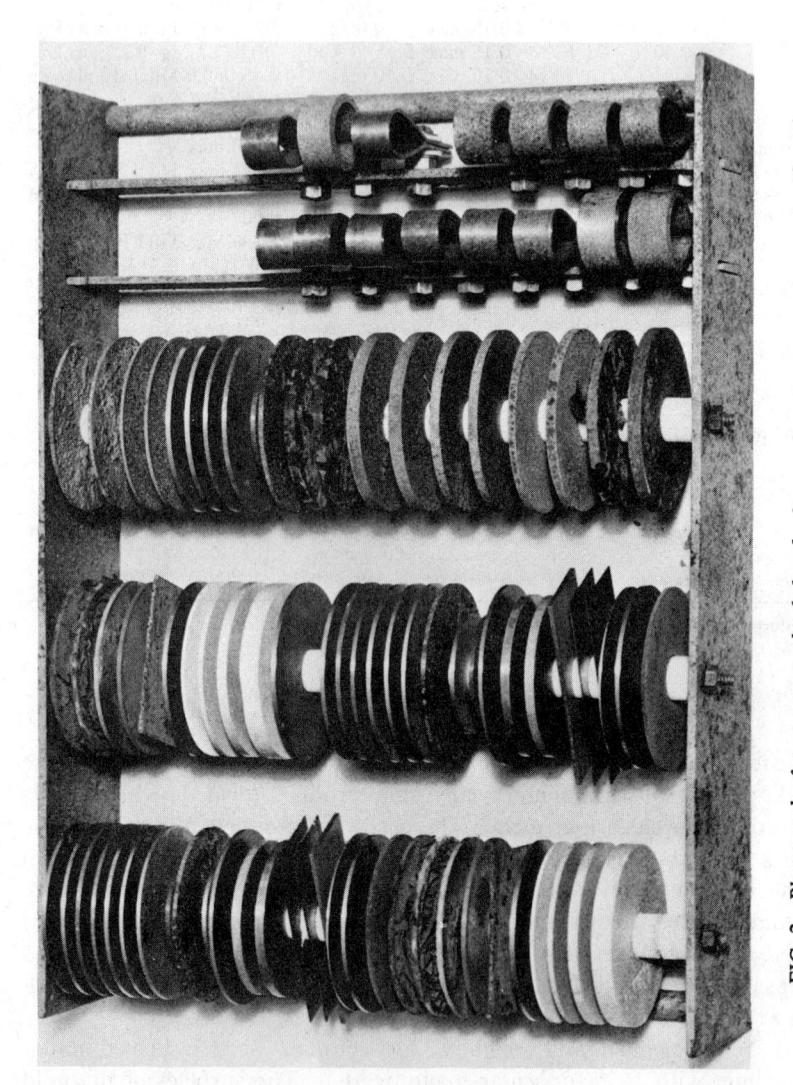

FIG. 2—*Photograph of coupons on a loaded rack after exposure in separated steam for 52 weeks.*

TABLE 4a—*The chemical cleaning methods used for removal of geothermal corrosion products.*[a]

Material	Cleaning Method[b]
Carbon steel, low-alloy steel	use A or B or both
Cast irons	use A only
Types 304, 321, 347, 316, 410 stainless steels; E-Brite 26-1, FV520 (B) 3RE60, ?RK65, Carpenter 20-Cb3, Incoloy 825	use C or B followed by D (and E if necessary)
Deoxidized copper and arsenical aluminum brass	use F and alternate with G after the major portion of soluble corrosion products have been removed.
Ti, IMI318, IMI550	use H
Nickel alloys, cobalt alloy, aluminum alloys, lead, zinc, tantalum, molybdenum, zinconum alloy GR32	under review
Removal of deposited amorphous silica	use I

[a] Methods A, B, C: See Refs 6 and 7; D, E: modifications of ASTM Recommended Practice G 1-72, see also Refs 8 and 9; G: see Ref 10.
[b] The cleaning methods are individually described in Table 4b.

TABLE 4b—*Description of individual chemical cleaning methods from Table 4a.*

Cleaning Method	
Method A	
Electrolytic cleaning in:	
Triammonium citrate	200 g
Distilled water to make	1 litre
Temperature	70°C (158°F)
Time (10 min, repeated)	10 to 40 min
Anode	carbon
Cathode	test material
Current density	0.1 to 2 A/dm^2
Method B	
Electrolytic cleaning in:	
Dihydrogen ammonium citrate made from	
Triammonium citrate	233 g
Citric acid	402 g
Distilled water to make	3 litres
Solution pH	3
Temperature	70°C (158°F)
Time (10 min, repeated)	10 to 40 min
Anode	carbon
Cathode	test material
Current density	0.1 to 2 A/dm^2
Method C	
Dip in:	
Triammonium citrate	200 g
Distilled water to make	1 litre

TABLE 4b—continued.

Cleaning Method	
Temperature	60°C (140°F)
Time (10 min, repeated)	10 to 40 min
Method D	
Dip in:	
Nitric acid (HNO$_3$, specific gravity 1.42)	200 ml
Distilled water to make	1 litre
Temperature	60°C (140°F)
Time (5 to 10 min, repeated)	5 to 20 min
Timing control—by indication of hydrophobic surface film in acid solution, a modification of the ASTM Test for Hydrophobic Surface Films by the Water-Break Test [F 22-65(1976)].	
Method E	
Electrolytic cleaning in:	
Hydrochloric acid (HCl, specific gravity 1.19)	0.5 litres
Hexamine	2.0 g
Distilled water to make	1 litre
Temperature	room
Time	5 to 20 min
Anode	carbon
Cathode	test material
Current density	1 to 2 A/dm^2
Method F	
Dip in:	
Potassium cyanide	50 g
Distilled water to make	1 litre
Temperature	room
Time	1 to 2 min
Method G	
Dip in:	
Sulphuric acid (H$_2$SO$_4$, specific gravity 1.25)	100 ml
Distilled water to make	1 litre
Temperature	room
Time	1 to 10 min
Method H	
Wash in cold water and scrub lightly using a medium hard nylon bristle brush. Remove all foreign corrosion products and monitor weight again.	
Method I	
Electrolytic cleaning in:	
Sodium hydroxide	100 g
Distilled water to make	1 litre
Temperature	70°C (158°F)
Time (10 min, repeated)	10 to 40 min
Anode	carbon
Cathode	test material
Current density	1 to 2 A/dm^2

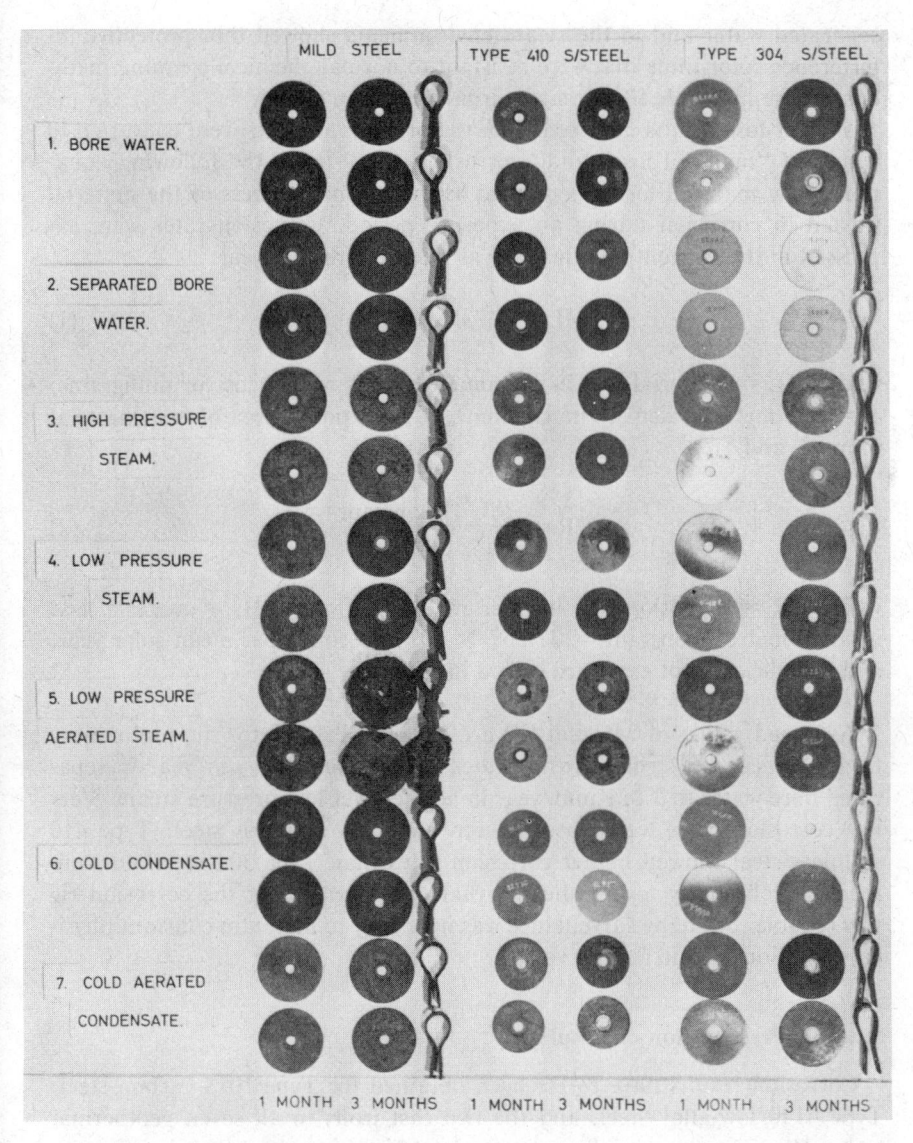

FIG. 3—*Coupons after exposure in preliminary trials.*

only shallow pitting. Corrosion pits were observed on many of the Type 304 stainless steel coupons, but these were shallow and few in number. Similar results were observed on the Type 410 stainless steels, except those exposed to separated steam (maximum pit depth, 0.14 mm), wet low-pressure aerated steam (maximum pit depth, 0.16 mm), and aerated condensate (maximum pit depth, 0.25 mm). Type 410 stainless steel coupons exposed in separated water and in the steam environments showed thin protective interference color films that were resistant to normal chemical cleaning methods. These insoluble films may be iron-chromium spinel.

Weight-loss results obtained from the preliminary tests were converted to values of "material loss" and "corrosion rate," using the following equations. The material loss was defined as the mean thickness of the material wasted in corrosion during an exposure period. Corrosion rates were expressed in the conventional manner as micrometres per year.

$$ML = (\delta/\rho A) \times 10^{+3} \qquad (1)$$

where ML = material loss in micrometres, δ = weight loss in milligrams, ρ = test material density in mg/mm^3, A = exposed area of test material in mm^2, and

$$CR = \frac{ML \times \text{constant}}{t} \qquad (2)$$

where CR = corrosion rate in micrometres per year, ML = material loss in micrometres, constant = 31.557×10^6 in seconds per mean solar year, and t = duration of exposure period in seconds.

Averaged values of ML and CR are given in Table 5 for the preliminary tests. The carbon steel corrosion rates varied from 2.64 μm/year in separated bore water to 3.321 mm/year in aerated wet low-pressure steam. Very low corrosion rates were observed for Type 304 stainless steel. Type 410 stainless steel showed higher corrosion rates than Type 304 in all environments. Preliminary tests indicated that the operation of the corrosion rig was reliable, but daily surveillance was necessary to maintain constant physical conditions within the test vessels.

Graphic Presentation of Results

Corrosion results have so far been obtained for Type 1018 carbon steel, Type 4140 low-alloy steel, and the two cast irons in all seven geothermal test media. These materials were examined first because of their importance in the construction of geothermal power plants. Other materials were also evaluated after exposure to wet low-pressure steam and to aerated cold condensate. These included ferrous alloys, titanium alloys, copper, and aluminum brass. The weight-loss measurements were supplemented by X-ray analysis of the corrosion products.

Results of these tests are reproduced in Figs. 4, 5, 6, and 7. The logarithmic plots of material loss (thickness) against exposure time shown in these figures give a simple empirical method of comparing corrosion results

TABLE 5—*Coupon weight-loss results for preliminary trials.*[a]

Environment		Carbon Steel Exposure Time		Type 410 Stainless Steel Exposure Time		Type 304 Stainless Steel Exposure Time	
		4 Weeks	13 Weeks	4 Weeks	13 Weeks	4 Weeks	13 Weeks
Bore fluid	ML	0.956	2.08	0.230	0.202	0.0185	0.0441
	CR	12.52	8.37	3.01	0.810	0.242	0.177
Separated water	ML	0.569	0.657	0.110	?	0.410	0.438
	CR	7.46	2.64	1.45	?	0.537	0.176
Separated dry steam	ML	8.67	25.50	0.323	1.22	0.0511	0.066
	CR	113.6	102.4	4.23	4.91	0.669	0.265
Wet low-pressure steam	ML	5.91	14.99	0.234	0.090	0.0256	0.0294
	CR	77.4	60.2	3.07	0.362	0.335	0.118
Aerated wet low-pressure steam	ML	72.9	802.9	0.628	5.30	0.0388	0.0488
	CR	955	3321	8.23	21.3	0.508	0.196
Condensate	ML	9.70	26.79	0.070	0.096	0.0321	0.0379
	CR	127.1	107.6	0.916	0.386	0.421	0.152
Aerated condensate	ML	197.0	446.4	0.064	0.130	0.0337	0.0235
	CR	2581	1793	0.843	0.524	0.441	0.0945

[a] Abbreviations used:
ML = material loss (μm).
CR = corrosion rate (μm/year).

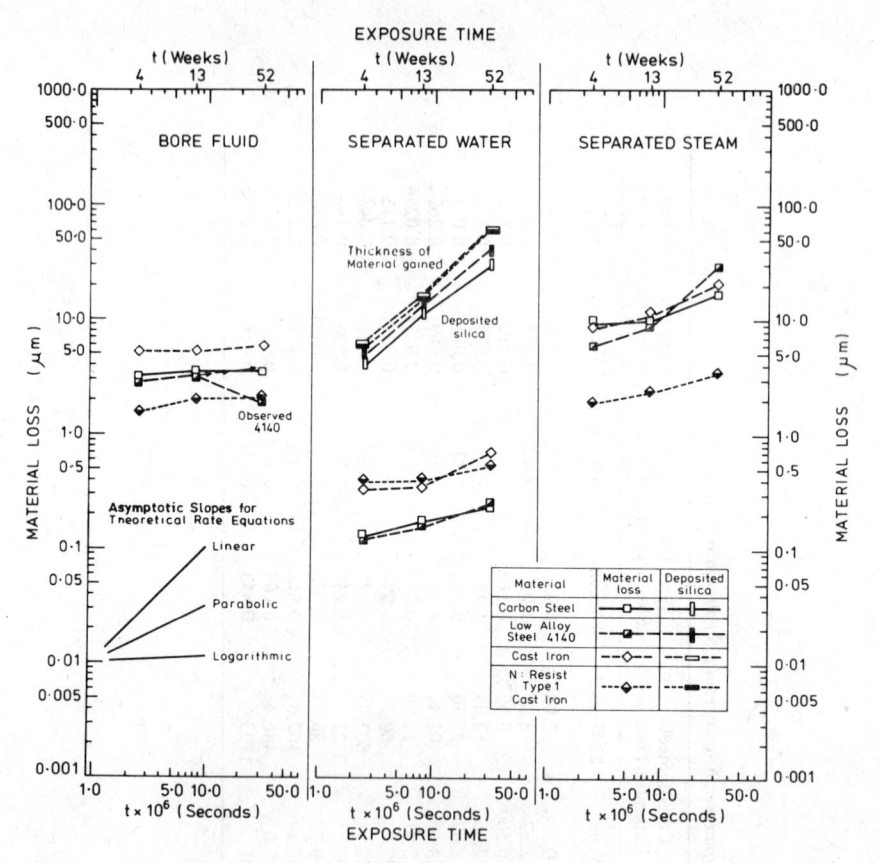

FIG. 4—*Logarithmic representation of material-loss results for selected ferrous alloys in bore fluid, separated water, and separated steam.*

for various materials with theoretical equations for film formation [11,12]. The slope should be unity for linear kinetics and 0.5 for a parabolic law, under diffusion control. In a strict sense there is no linear portion for logarithmic growth, but, for times of the order of those used, the curve approximates a straight line with a slope of 0.06. The use of material loss as a criterion is necessary because the weight-gain results required for direct comparison with theoretical film-formation kinetics are generally not available. This is because in extended geothermal field tests multiple-layer films may form, corrosion products often flake off, or foreign materials may be entrained in the products.

Although the observed slopes may not dictate precise growth mechanisms, they do provide useful indications of these mechanisms. The limited number of replicate specimens prohibited statistical manipulation of the data.

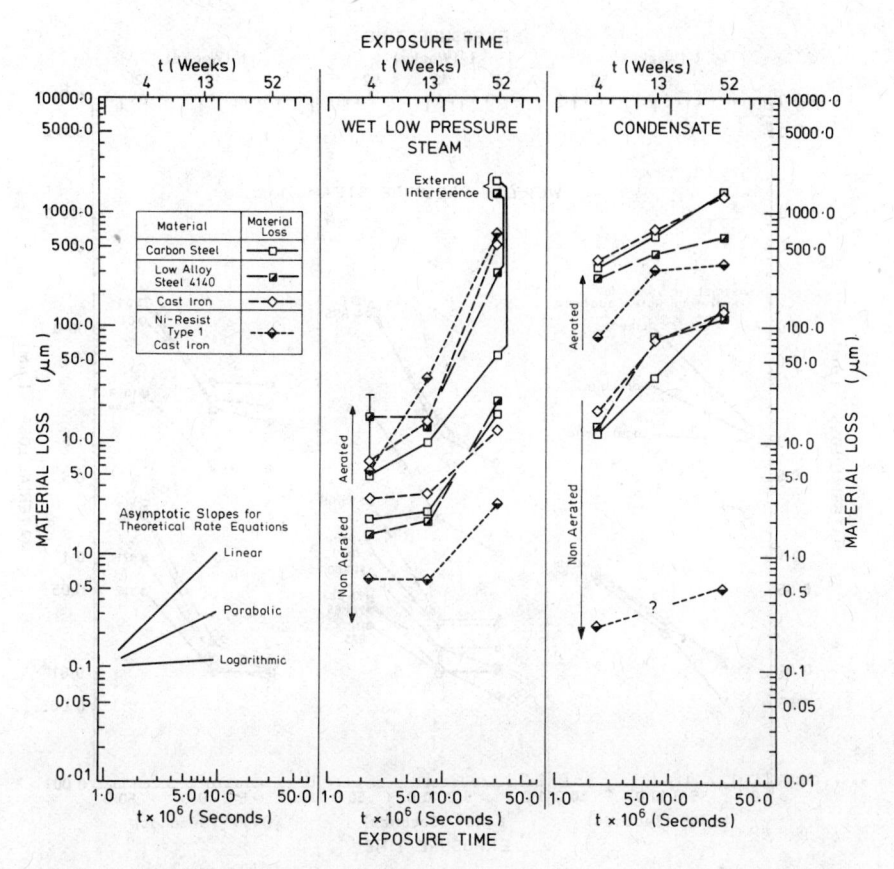

FIG. 5—*Logarithmic representation of material-loss results for selected ferrous alloys in wet low-pressure steam and condensate (nonaerated and aerated).*

Corrosion Characteristics of Carbon Steel, Low-Alloy Steel, Gray Cast Iron, and Ni-Resist Type 1 Cast Iron

Corrosion Product Stability

X-ray diffractometer results of corrosion products collected from the carbon steel, the low-alloy steel, and the two cast irons in all of the test environments are summarized in Table 6. Minor amounts of other corrosion products would not be detected by this technique. The stable corrosion products in nonaerated environments were primarily iron sulfides. The stability limits of these phases have been predicted in the form of potential-pH diagrams for downhole conditions in the Broadlands field at 250°C and 4 MPa [13] and for separated steam at the selected test conditions of 160°C and 650 kPa (see Fig. 8) [14]. Mackinawite, pyrrhotite, and troilite were

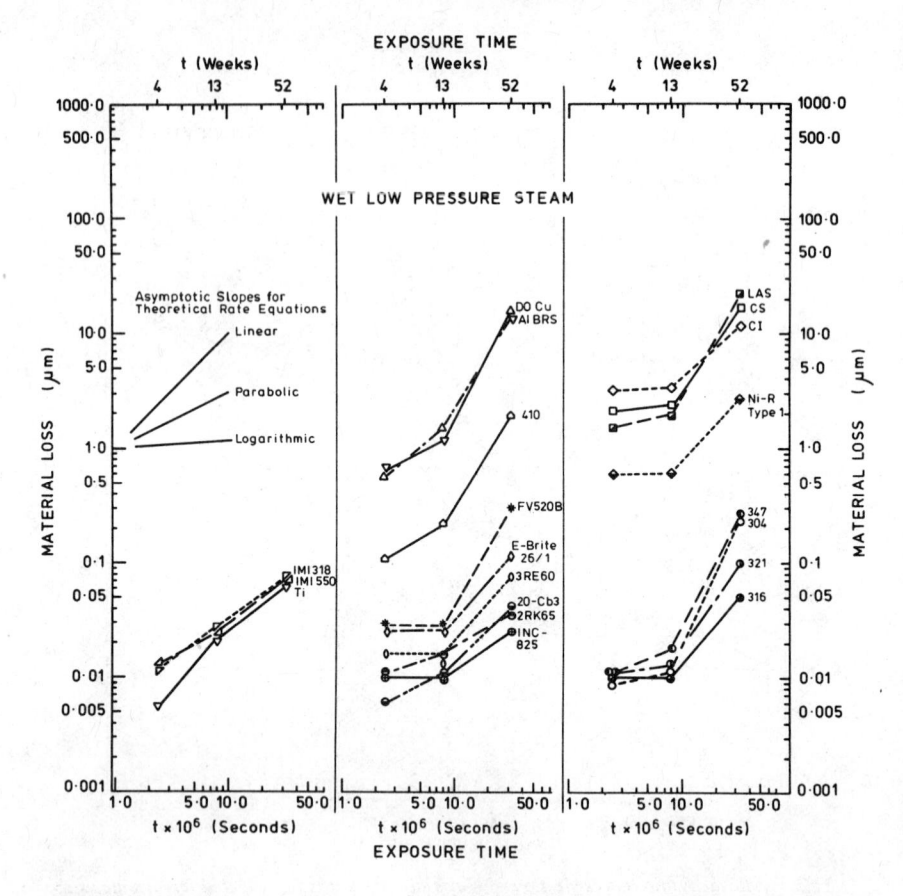

FIG. 6—*Logarithmic representation of material-loss results for selected alloys in nonaerated wet low-pressure steam.*

present in the separated steam test environment (Table 6), and Fig. 8 shows a stability region for ferrous sulfide (FeS) which suggests an oxidation potential range of −0.46 to −0.70 V at the measured pH value of 5.7. Corrosion product adhesion properties (macro effects) and the observed incidence of pitting on carbon and low-alloy steels and the two cast irons have been tabulated in Table 7. A more extensive evaluation of similar results for stainless steel alloys exposed for 52 weeks in wet low-pressure steam is given in Table 8.

Bore Fluid

Troilite and pyrite were the major corrosion products in bore fluid, although other unidentified diffracting materials were also observed (Table 6).

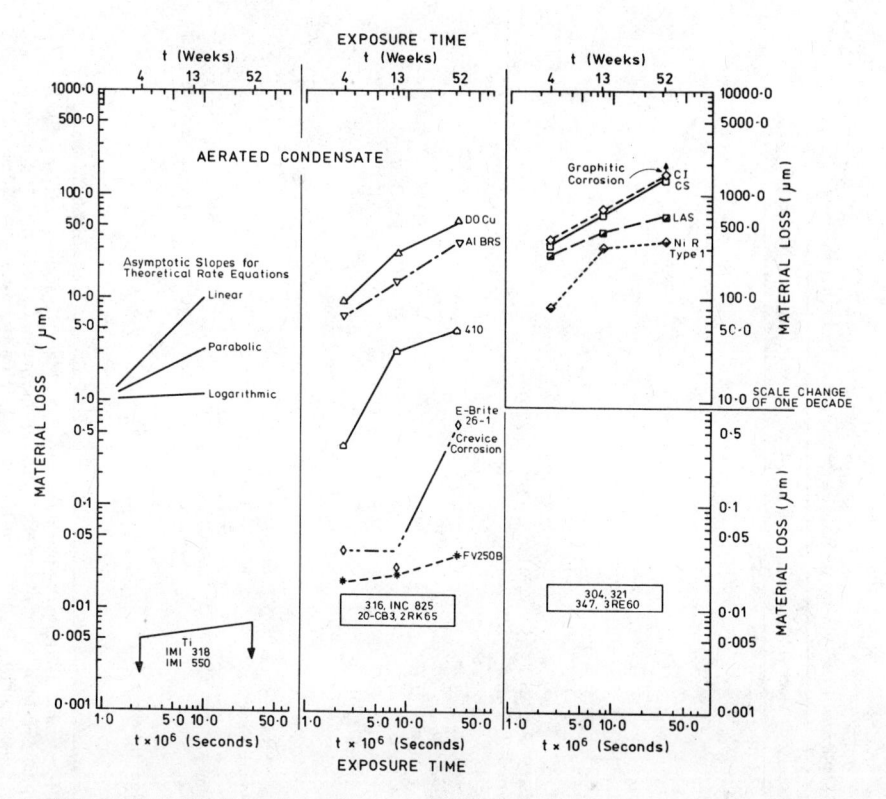

FIG. 7—*Logarithmic representation of material-loss results for selected alloys in aerated condensate.*

Good corrosion product adhesion was evident on all the materials, with the exception of some low-alloy steel coupons. The pitting of coupons exposed to bore fluid was generally confined to the region covered by the PTFE insulating spacers. Pits on the low-alloy steel were observed only after long exposure time, and, although the carbon steel coupons pitted sooner, the pits were shallow and had little effect on material losses.

Separated Bore Water

Material losses in separated bore water were much lower than those in bore fluid (Fig. 4). Stable corrosion products on carbon steel exposed in the preliminary experiments were identified as mackinawite and troilite, and deposits of α-quartz (a small amount) and amorphous silica were also confirmed. Chemical analysis of the deposits indicated the presence of 76 weight percent amorphous silica, with small amounts of aluminum oxide (Al_2O_3), sodium, potassium, and iron, identified by atomic absorption.

TABLE 6—Summary of corrosion products on some ferrous alloys identified by X-ray diffractometer analysis using Cu Kα radiation.[a]

Environment	Time, weeks	Carbon Steel 1018	Low Alloy Steel 4140	Cast Iron Gray	Cast Iron Ni-Resist Type 1
Bore fluid	4	T	...	...	...
	13	T + U	T	T + U	T + P + U
	52	AS	T + P	AS	AS
Separated bore water	4 + 13 + 52	AS + α-Quartz + Mac + T	AS		
Separated high-pressure steam	4	Mac + Pyrr	Mac + Pyrr + T + U	...	...
	13	Mac + Pyrr	Mac + Pyrr + T + U	...	...
	52	Mac + Pyrr	T	T	...
Wet low-pressure steam	4	Mac + U	...	Mac + U	...
	13	Mac + T	Mac + Pyrr + T	Mac + Pyrr	...
	52	Mac + T + P + Mgn + S + FeSO$_4$·H$_2$O	Pyrr + T + iron	T + P + Mar	...
Wet low-pressure aerated steam	4	P + H + C	P + Mgn + H + G + U	P + Mgn + H + C	...
	13	P + Mgn + H + C	P + Mgn + H + G + C	P + Mgn + H + G + C	S + Mgn + H + G + C + FeSO$_4$·H$_2$O
	52	P + T + H + G + C + FeSO$_4$·H$_2$O	S + Mgn + H + C + FeSO$_4$·H$_2$O	P + Mgn + H + G + C + S + Cem + FeSO$_4$·H$_2$O	Mgn + H + G + Sid + U
Condensate	[b]	S + FeSO$_4$·H$_2$O			
	4	Mac + Cem	Mac + Cem + U	Mac + Cem + C	...
	13	Mac + Cem	Mac + Cem	Mac + Pyrr + Cem + C + Pyrr crystals	...
	52				
Aerated condensate	4	...	Cem + Sid + U	S + Cem + Sid + C + U	G + C
	13	Mgn + Cem + Sid	...	...	V + C + U
	52	Mar + Mac + S + G + U	Mgn + Cem + Sid	...	V + Cha
	[c]				

[a] Abbreviations used: P = pyrite (FeS$_2$); Mar = marcasite (FeS$_2$); T = troilite (FeS); Pyrr = pyrrhotite (Fe$_{(1-x)}$S); Mac = mackinawite (Fe$_{(1+x)}$S); Cha = chalcopyrite (CuFeS$_2$); S = sulphur (S); Mgn = magnetite (Fe$_3$O$_4$); H = haematite (α-Fe$_2$O$_3$); G = geothite (α-Fe$_2$O$_3$·H$_2$O); Sid = siderite (FeCO$_3$); Cem = cementite (Fe$_3$C); C = graphite (C); AS = amorphous silica ((SiO$_2$) × H$_2$O); U = unknown.
V = violarite ((FeNi)$_3$S$_4$);

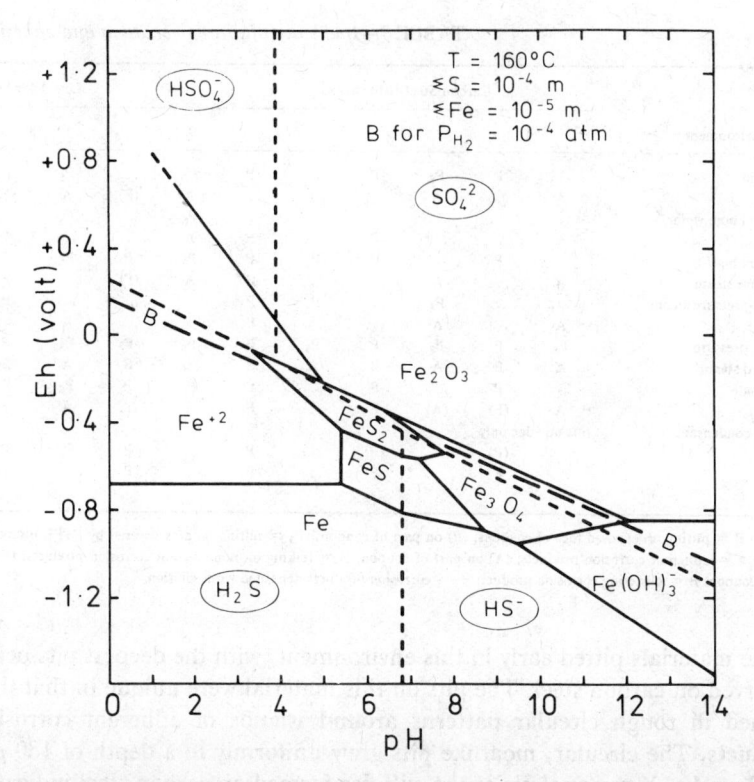

FIG. 8—*Potential-pH diagram for the system Fe-S-H$_2$O at 650 kPa at Broadlands* [14].

The thickness of deposited silica was calculated from weight-gain observations. The results of these calculations are shown in Fig. 4 as material gain (deposited silica) versus exposure time for the separated bore water environment. A maximum rate of deposition of 63.5 μm/year was observed, with a log-log slope of 0.9, indicating (within experimental accuracy) that a linear deposition rate of silicon dioxide (SiO$_2$) was occurring.

Although pitting was observed in the region covered by the PTFE insulating spacers on the low-alloy steel, the gray cast iron, and the Ni-Resist Type 1 cast iron, the pits were of small diameter, very shallow, and few in number.

Separated Steam

For short exposures in separated steam the sulfide films of mackinawite and pyrrhotite that formed were generally thin and adherent. Extensive flaking was observed after extended exposures, and troilite, or thicker scales of mackinawite and pyrrhotite, were observed. This flaking was most pronounced on the low-alloy steel, even after the initial 4-week exposure. All

102 GEOTHERMAL SCALING AND CORROSION

TABLE 7—*Incidence of pitting corrosion and adhesive*

	Carbon Steel 1018, weeks						Low-Alloy Steel	
Environment	4		13		52		4	13
Bore fluid	...	P	P_S	P	P_S	P	...	...	P_S	...
	A	...	A	...	A	...	A	(F)	A	(F)
Separated bore water[b]	...	...	...	...	...	...	P_S	...	P_S	...
	A	...	...	F	...	F	A	...	...	F
Separated high-	...	P	...	P	P_S	P	P_S	P	P_S	P
pressure steam	A	...	A	...	...	F	A	(F)	...	F
Wet low-pressure steam	...	...	P_S	...	P_S	P	...	...	...	P
	A	...	A	...	...	F	A	...	A	...
Wet low-pressure	P_S	P	P_S	P	P_S	P	P_S	P	P_S	P
aerated steam	A	B	A	B	A	B	A	B	A	B
Condensate	...	P	...	P	...	P	P_S	P	P_S	P
	A	(F)	(A)	F	...	F	A	...	A	...
Aerated condensate	pits on edge only									
	...	(P)	...	(P)	...	P	...	P	P_S	P
	...	F	...	F	...	F	...	F	...	F

[a] Key: P = pitting on exposed face of coupons, (P) on part of coupon; P_S = pitting on area covered by PTFE insulating spacers; A = adherent corrosion products; (A) on part of coupon; F = flaking on nonadherent corrosion products; (F) on part of coupon; B = blistering corrosion product; + = corrosion products deposited from solution.

of the materials pitted early in this environment, with the deepest pits being observed on carbon steel. The pits on this material were unique in that they formed in rough circular patterns around islands of adherent corrosion products. The circular, moatlike pits grew uniformly to a depth of 180 μm in 52 weeks. X-ray analysis of the sulfides formed on carbon steel indicated the formation of increasing amounts of pyrrhotite close to the iron-sulfide interface.

Wet Low-Pressure Steam

Iron sulfides formed the bulk of the stable corrosion products in unaerated wet low-pressure steam (Table 6). There was a distinct change in corrosion behavior after the 13-week exposure, as shown in Fig. 5. The 52-week exposures show material losses that are five to ten times higher than would be expected if logarithmic or parabolic kinetics were occurring. The cleaned coupons exposed for 52 weeks revealed extensive pitting, whereas little or no pitting was observed on those exposed for the shorter periods. The onset of pitting and poor adhesion of corrosion products on coupons of carbon and low-alloy steels (Table 6) give a reason for the change in corrosion rates. On carbon steel, the products also changed with increasing exposure time from mackinawite to mackinawite plus pyrrhotite and troilite. Where active pitting had occurred, the scale contained mackinawite, troilite, and pyrite, plus a small amount of elemental sulfur, and hydrated ferrous sulfate.

properties of corrosion products on some ferrous alloys.[a]

4140, weeks		Cast Iron—Gray, weeks						Cast Iron Ni-Resist Type 1, weeks					
52		4		13		52		4		13		52	
P$_S$	P	...	...	P$_S$	...	P$_S$	...	P$_S$	...	P$_S$	...	P$_S$	...
A	(F)	A	(F)	A	...	A	...	A	...	A	...	A	...
P$_S$	...	P$_S$	...	P$_S$	...	P$_S$	...	P$_S$	...	P$_S$	...	P$_S$	...
...	F	A	...	A	...	...	F	A	...	...	F	...	F
P$_S$	P	P$_S$	P	P$_S$	P	P$_S$	P	P$_S$	...	P$_S$	...	P$_S$	...
...	F	A	(F)	A	(F)	...	F	A	...	A	...	A	...
P$_S$	P	...	...	...	...	P	P$_S$	P	...	...	...	P$_S$	P
...	F	A	...	A	...	A	...	A	...	A	...	A	...
P$_S$	P	P$_S$	P	P$_S$	P	P$_S$	P	P$_S$	P	P$_S$	P	graphite shell	
A	B	A	B	A	B	A	B	A	B	A	B	A	B
P$_S$	P	P$_S$	?	P$_S$	?	P$_S$	?	P$_S$	P	P$_S$	P	P$_S$	P
A	...	A	...	A	...	A	...	A	...	A	...	A	...
		rough surface				graphite shell				rough surface			
P$_S$	P	...	(P)	P$_S$	(P)			P$_S$	(P)	P$_S$	(P)	P$_S$	(P)
...	F	A	...	A	...	A	...	A	...	A	...	A	...
+		+		+		+		+		+		+	

[b] A and F in separated water refer to adhesion of deposited silica.

Wet Low-Pressure Aerated Steam

Aeration of the wet low-pressure steam resulted in stable corrosion products, which included iron oxides (hematite, geothite, and magnetite), as well as iron sulfides (troilite and pyrite), elemental sulfur, and hydrated ferrous sulfate. Blistering of corrosion products was observed after short exposures, and, after 52 weeks, thick, porous, blistered, but adherent, corrosion products were formed. The corrosion kinetics are demonstrated in the log-log plots of material loss versus time shown in Fig. 5. Exceptional weight losses were obtained for two exposed coupons, a carbon and a low-alloy steel exposed for 52 weeks, but the required electrical isolation of these two coupons was not maintained. Graphitic corrosion was evident on the two cast irons. The Ni-Resist Type 1 cast iron performed very poorly in the aerated environment, in direct contrast to the results in nonaerated steam and bore fluid. All of the coupons indicated severe nonuniform corrosion. Large round pits were evident on carbon steel, whereas high-density smaller-diameter pits were formed on low-alloy steel. These differences may be due to previously noted structural differences (Table 3a). Large shallow pits were again noted on the gray cast iron, but the Ni-Resist Type 1 cast iron had general uneven corrosion with a high surface roughness. Coupons of this material exposed for 52 weeks consisted of a shell of graphite with a tough central core of porous iron. None of the coupons was perforated despite the formation of deep pits and high material losses.

TABLE 8—*Classification of pitting on stainless steels exposed for 52 weeks in wet low-pressure steam.*

Exposed Material Designation	Number of Specimens	Pit Density, ASTM Designation (G 46–76)	On Face,[b] μm Avg	(Max)	At Edge of Spacer,[c] μm Avg	(Max)	Under Spacer,[c] μm Avg	(Max)
S41000	3	1 (13 weeks)	13	(23)	28	(49)	24	(40)
S30400	2	>5	33	(67)	38	(62)	21	(26)
FV520 (B)	3	5 to >5	32	(67)	35	(74)	25	(72)
S34700	2	5	18	(52)	29	(42)	14	(24)
E-Brite 26-1	2	4	36	(59)	43	(68)	29	(55)
S32100	2	4	24	(38)	36	(63)	20	(43)
3RE60	2	4	21	(53)	27	(43)	15	(32)
S31600	2	3 to 5	17	(42)	19	(28)	18	(26)
NO8825 Incoloy 825	3	3 to 4	23	(49)	31	(57)	25	(37)
2RK65	3	2 to 3	16	(48)	24	(48)	13	(21)
NO8020 Carpenter 20Cb3	3	1	11	(55)	19	(25)	18	(40)
			15	(28)	16	(27)	11	(23)

[a] The pit sizes were all <0.5 mm². The pits showing the greatest depth were selected for measurement.
[b] 16 to 24 readings/exposed material.
[c] 12 to 18 readings/exposed material.

Condensate

In nonaerated condensate, mackinawite and pyrrhotite were the dominant stable iron sulfides. Crystals of pyrrhotite were observed on the cast iron coupons exposed for 52 weeks, and siderite was identified on low-alloy steel. The Ni-Resist Type 1 cast iron showed outstandingly low material loss in the condensate, but the results may be somewhat low because of the difficulty of ensuring complete chemical cleaning of this material. The corrosion of carbon steel followed linear kinetics, whereas the low-alloy steel and gray cast iron showed a distinct change from high to low corrosion rates between the 13 and 52-week exposures. Pitting was common to all the materials but was most pronounced on carbon steel and low-alloy steel and preferentially followed the grain orientation, which was perpendicular to the exposed faces of the low-alloy steel coupons and parallel in the case of carbon steel. On the low-alloy steel a honeycomb pattern of pits formed early in the exposure and increased in depth with increasing exposure time. The effect of the deep narrow pits was to provide good adhesion of the corrosion products to the underlying steel, and it is this which caused the slowing of material loss with time. By contrast, the lack of adhesion of corresponding products on carbon steel resulted in continuous uniform attack. Pitting on the exposed faces of the carbon steel coupons was associated with hydrogen blistering and was distributed in lines along the rolling direction. This pitting was of much lower density than that on the low-alloy steel. The pitting on the exposed end grain of the carbon steel coupons was similar to that noted on the faces of the low-alloy steel.

Aerated Condensate

Material losses observed on these materials increased substantially on aeration of the condensate. The corrosion products formed in this environment were often difficult to type because of oxidation on drying in air. Iron oxides, iron sulfides, iron carbonate, sulfur, and a variety of other corrosion products peculiar to particular alloys were identified (see Table 6). The corrosion products were extremely nonadherent on both carbon and low-alloy steel; however, the low-alloy steel demonstrated superior corrosion resistance (Fig. 5). Gray cast iron lost material by graphitic corrosion, and the 52-week specimens had no residual iron. The Ni-Resist Type 1 cast iron performed well in this environment, possibly because of the formation of violarite and chalcopyrite (Table 6). The generally rough surface observed after cleaning, and the identification of graphite in the corrosion products does, however, suggest that this alloy may be liable to graphitic corrosion in this environment. Some foreign corrosion products were observed on these materials, but these products were not considered to have affected the corrosion characteristics of the materials. The material losses were generally uniform, with only a few shallow pits being observed.

Stainless Steels, Copper Alloys, and Titanium Alloys

Wet Low-Pressure Steam

The corrosion properties of stainless steels and the nickel-iron-chromium alloy Incoloy 825 in wet low-pressure steam are summarized in Fig. 6. The corrosion properties of carbon and low-alloy steel and cast irons, as well as those of deoxidized copper, aluminum brass, and titanium alloys, are also shown for comparison. The majority of the stainless steels had low initial material losses, followed by intensified corrosion after 13 weeks. The coupons exposed for 4 and 13 weeks had thin surface films, giving a tarnished appearance, but large numbers of localized defects were evident on those exposed for 52 weeks, as illustrated in Figs. 9 and 10. Visual inspection after cleaning confirmed the onset of pitting corrosion at some time after 13 weeks. Pit densities and depths on the stainless steel coupons exposed for 52 weeks were measured by the ASTM Recommended Practice for Examination and Evaluation of Pitting Corrosion (G 46-76). A hierarchy based on pit density was established, and measurements of pit depths provided additional information for ranking materials with similar pit densities (Table 8). The pits on all of the stainless steels were less than 0.5 mm^2 in area, and the maximum pit depths varied from 21 to 74 μm.

Deoxidized copper and aluminum brass were uniformly corroded in this environment. The aluminum brass formed adherent films, whereas those formed on deoxidized copper flaked off readily after the 52-week exposure. On deoxidized copper α-chalcocite (Cu$_2$S) was identified. Digenite (Cu$_9$S$_5$), zinc oxide (ZnO), and wurtzite (ZnS) were stable corrosion products found on aluminum brass.

Removal of the corrosion products from the titanium alloys proved an intractable problem. Thin films showing interference color fringes were, however, readily observed, and the exposed coupons registered weight gain that increased with greater exposure time. The results shown for titanium alloys are based on observations of weight gain assumed to be due to the formation of rutile (TiO$_2$), and material losses were calculated by a method based on the ratio of molecular volumes of rutile and titanium. The corrosion products on the titanium alloys have not been identified. An independent study on the interpretation of oxide colors on titanium has been done [15]. Good correlation has been observed between the oxide thickness calculated from weight gain results and the oxide thickness indicated by interference color fringes. No pitting was evident on the titanium alloys, and the formed films were protective to the underlying material (Fig. 6).

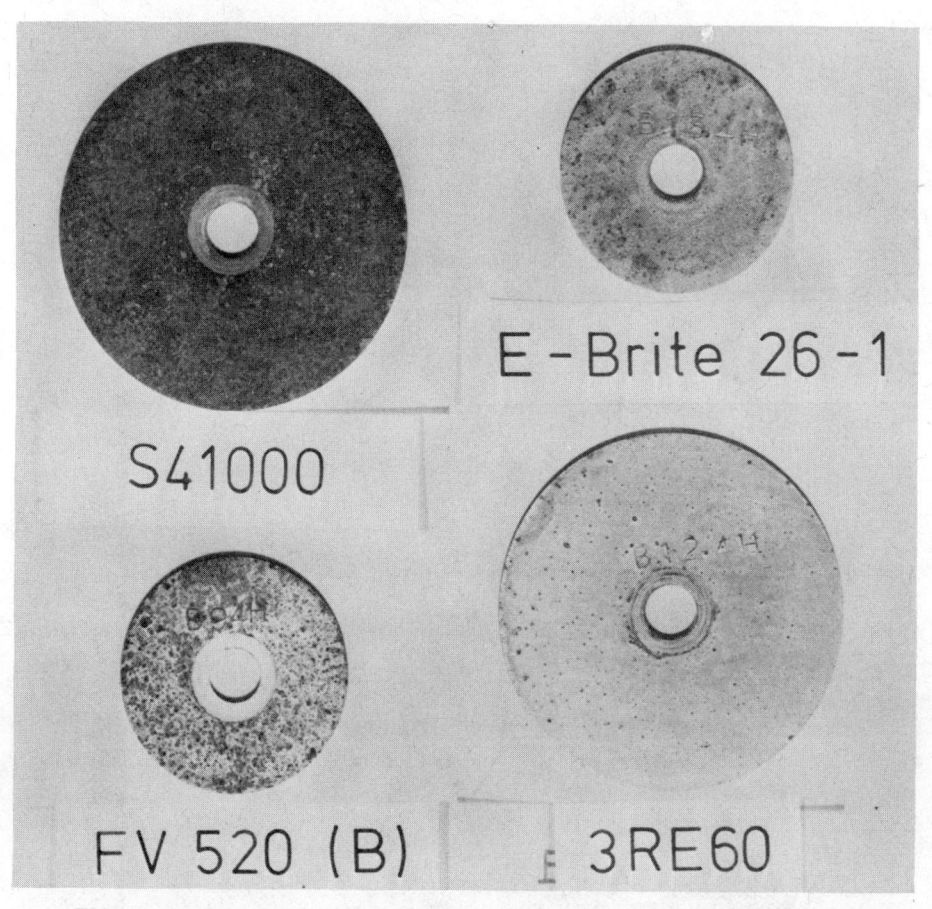

FIG. 9—*Typical appearance of stainless steel coupons exposed for 52 weeks in wet low-pressure steam.*

Aerated Condensate

The corrosion properties of the stainless steels in aerated condensate are summarized in Fig. 7. The heavily corroded ferrous alloys and some nonferrous alloys are also shown for comparison. Except for Type 410, the stainless steels and the titanium alloys were covered with thick adherent films of unidentifiable rust-colored iron corrosion products deposited from solution. The high iron content of the solution was due to corrosion of the carbon steel pressure vessel and the more corrodible ferrous alloys. Very low material losses were obtained for the majority of the stainless steels, and it was convenient to group alloys with similar results, as shown in Fig. 7.

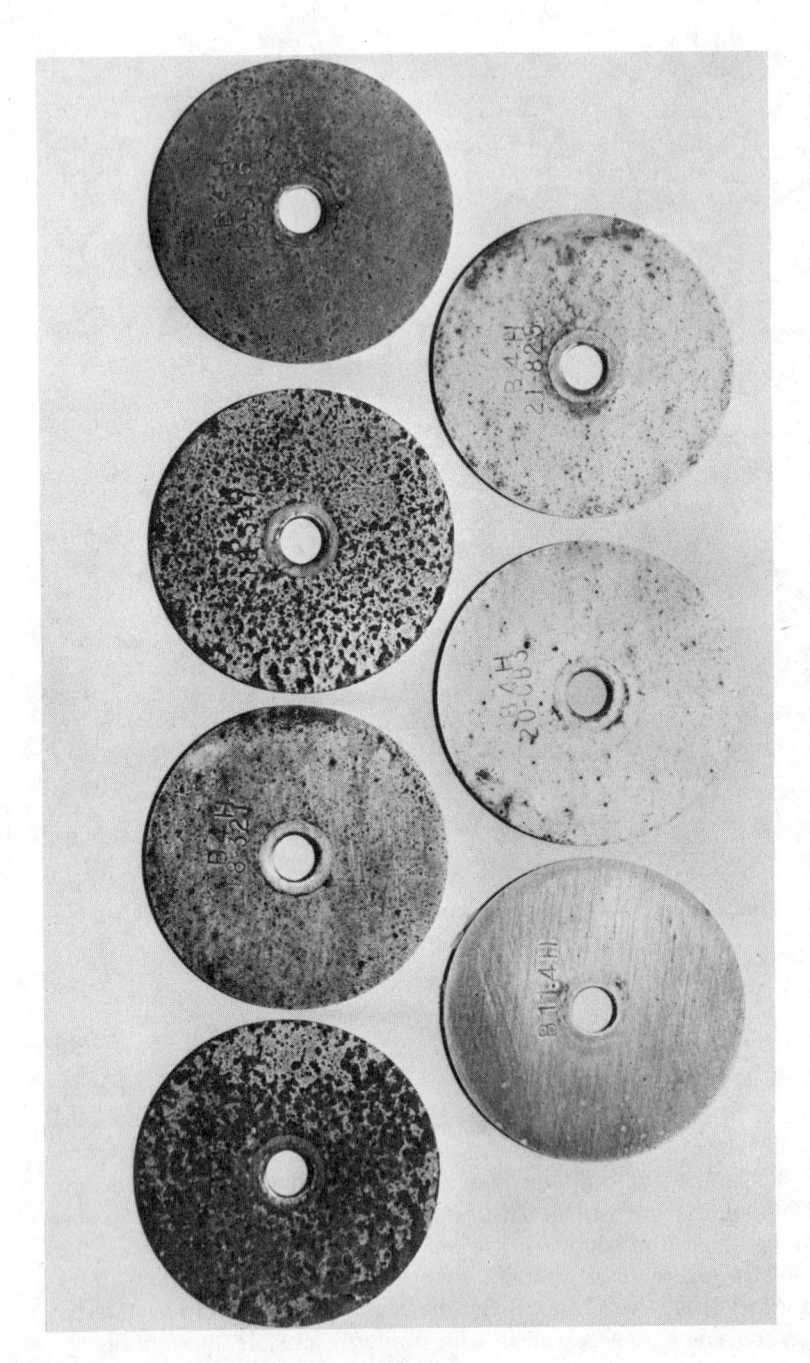

FIG. 10—*Typical appearance of stainless steel coupons exposed for 52 weeks in wet low-pressure steam. Top row: S30400, S32100, S34700, S31600; bottom row: Sandvik 2RK65, Carpenter 20Cb3, Incoloy 825.*

The alloy E-Brite 26-1 and the Type 410 stainless steels showed significantly higher material losses. The material loss of E-Brite 26-1 at the 52-week exposure period was due to the formation of large-diameter pits located under the PTFE insulating spacer. The Type 410 stainless steel pitted during the shortest exposure time, and, although film formation was evident, the major material loss was due to the growth of large-diameter deep pits. Pitting was evident on the stainless steels with low material losses, but the pits were very shallow and few in number.

Deoxidized copper and aluminum brass formed corrosion products that combined with the iron compounds deposited from solution, and the resultant composite films had poor adhesion. Only the copper-bearing corrosion products remained in contact with the aluminum brass coupons—copper sulfide (Cu_xS, $1.96 > X > 1.86$) and digenite. Cuprite (Cu_2O), digenite, and copper sulfide (Cu_xS, $1.96 > X > 1.86$) were observed on the deoxidized copper. Both of these alloys had corroded uniformly.

Weight losses were detected for the titanium alloys after deposited corrosion products had been removed by the dihydrogen ammonium citrate cleaning method (Table 4). The generalized representation of material loss reproduced in Fig. 7 was therefore calculated from weight losses. No pitting was observed on the titanium alloys.

Discussion of Results

Exposure of ferrous metal coupons in geothermal fluids have shown varying rates of material loss and differing corrosion products, which depend primarily on the environment and secondarily on the material. Changes in the surface film properties of adhesion and porosity, and localized pitting corrosion, coincide with these variations.

Protective films were formed on carbon steel, low-alloy steel, and the two cast irons in bore fluid, separated bore water, and separated steam. In separated water the corrosion of these ferrous alloys appeared to be severely inhibited by the deposition of amorphous silica from solution. Deposited silica was adherent during the exposure, and the observed continuing corrosion must have been by diffusion processes through these films. Corrosion proceeded more slowly in bore fluid, where troilite and pyrite were the major stable iron sulfides. In separated steam, mackinawite, pyrrhotite, and troilite were stable, and corrosion occurred at approximately parabolic rates. The higher initial material losses, increased tendency to pitting corrosion, and decreased film adhesion also suggest that the composite films, which included mackinawite, were less protective.

The increased corrosion in separated steam may be attributed to the preferential partitioning of CO_2 and H_2S gases into the steam phase and most of the ammonium (NH_4) into the separated water [16]. In addition, the dissolved silica and alkaline salts are concentrated in the separated water.

The poor performance of a low-alloy steel (slope, 0.5) compared with carbon steel (slope, 0.2) in separated steam is of importance because of the proposed use of similar low-alloy steels in power plant applications.

Initial exposures of ferrous alloys in wet low-pressure steam resulted in protective film formation and lower material losses than those in bore fluid and separated steam. The high corrosion observed after 52 weeks was attributed to metal pitting. The appearance of pyrite, magnetite, sulfur, and hydrated ferrous sulfate in the scales was also a feature of this exposure period. Although the cause of the initiation and growth of pits on these materials is not immediately obvious, the possibility of oxygen entering the test vessel cannot be discounted, especially since elemental sulfur and hydrated ferrous sulfate have been observed. Because the pressure is only 26 kPa above atmospheric, it would be possible for small concentrations of oxygen to enter the vessel by back diffusion or venturi effects through valve packing materials. The high wetness (10 percent) and lower temperature (104°C of this vessel, compared with separated steam at 160°C) make this environment more susceptible to oxygen contamination during specimen removals. The effects of low oxygen levels on corrosion rates and the chance of oxygen contamination are currently being investigated.

Aeration of the wet low-pressure steam led to rapid linear corrosion rates for carbon steel, low-alloy steel, and the two cast irons. The porosity of the complex corrosion products permits corrosive condensates to remain in close proximity to the metal surfaces, resulting in increasing corrosion rates with exposure time. The formation of iron oxides, sulfur, and hydrated ferrous sulfate, and the reduced stability of iron sulfides are clearly due to the high oxygen levels. The mechanisms for corrosion in these conditions are still speculative.

In condensates, a wide range of corrosion kinetics was observed. In nonaerated condensate mackinawite, scales were detected with linear corrosion kinetics and no protective films, except where the corrosion products were mechanically tied by formation of corrosion products in pits, which grew into the material along the rolling direction. Aeration increased the material losses considerably, and the corrosion followed a parabolic rate with some degree of protection provided by the scales of magnetite and siderite. Cast irons were susceptible to graphitic corrosion in aerated environments but performed well in nonaerated conditions.

The kinetic behavior of stainless steels in wet low-pressure steam was similar to that of the less alloyed ferrous materials. The pitting of these alloys, detected after 52 weeks of exposure, dominated the corrosion kinetics, and the incidence of pitting was directly reflected in the material loss results. The possibility of oxygen causing pitting corrosion in this environment, as previously discussed, is being investigated.

Type 410 stainless steel is of particular interest because of its use as a turbine blade material in geothermal power stations. Although only troilite and pyrite were positively identified on Type 410 stainless steel exposed in this environment, the coupons had a compact adherent interference color film close to the metal. In similar power plant applications where the alloy was subjected to pitting corrosion, Type 410 stainless steel formed iron chromium spinel layers on all surfaces, including the roots of potentially dangerous pits.

Films formed on titanium alloys were shown indirectly to form protective films—parabolic kinetics. All of the alloys exposed had similar corrosion behavior, so that selection for particular applications would be based on their mechanical properties. The immunity of these alloys to pitting corrosion was an important result of this investigation.

In aerated condensate, stainless steels followed logarithmic corrosion kinetics with the exception of E-Brite 26-1 and Type 410. The former was susceptible to crevice corrosion, and Type 410 stainless steel pitted readily in spite of obvious film-forming tendencies. The low corrosion of the remaining stainless steels and the titanium alloys indicates that the film-forming properties and corrosion resistance of these particular alloys were not adversely affected by the extraneous deposition of iron corrosion products from solution.

Conclusions

Results of corrosion tests on coupons exposed in a geothermal test rig have been presented graphically and show the following main features:

1. Protective films are formed on ferrous alloys in bore fluid and separated water and, to a lesser extent, on those exposed to separated steam.
2. The breakdown of protective films by pit formation was observed on ferrous alloys after extended exposures in wet low-pressure steam.
3. Nonprotective scales were formed on some ferrous alloys in condensate.
4. Corrosion was increased if air was added to the wet steam or condensate.
5. Ni-Resist Type 1 cast iron had good corrosion resistance in nonaerated steam environments and condensate.
6. Continuous thin films were formed on titanium alloys in wet low-pressure steam.

References

[1] Marshall, T. and Braithwaite, W. R., *Unesco, 1973. Geothermal Energy, Earth Sciences*, Vol. 12, pp. 151–160.
[2] Mahon, W. A. J. and Finlayson, J. B., *American Journal of Science*, Vol. 272, Jan. 1972, pp. 46–68.

[3] Lyle, O., *The Efficient Use of Steam*, His Majesty's Stationery Office, London, 1947, pp. 170-174.
[4] Mahon, W. A. J. and Finlayson, J. B., "The Geochemistry of the Broadlands Geothermal Area," Chemistry Division, Department of Scientific and Industrial Research, Wairakei, New Zealand, 1976.
[5] Glover, R. B., "The Chemistry of BR-22 Corrosion Rig Fluids," Report No. CD 118/12-RBG 33, Chemistry Division, Department of Scientific and Industrial Research, Wairakei, New Zealand, 1978.
[6] Blume, W. J. in *Cleaning Stainless Steel, ASTM STP 538*, American Society for Testing and Materials, Philadelphia, 1973, pp. 43-53.
[7] Blume, W. J., *Materials Performance*, Vol. 16, No. 3, March 1977, pp. 16-19.
[8] Lane, G. S. and Ellis, J., *Corrosion Science*, Vol. 11, 1971, pp. 661-663.
[9] MacMillan, J. W. M. and Flewett, P. E. J., *Micron*, Vol. 6, 1975, pp. 141-146.
[10] Ailor, W. H., *Handbook on Corrosion Testing and Evaluation*, Wiley, New York, 1971, pp. 133-138.
[11] Kubaschewski, O. and Hopkins, B. E., *Oxidation of Metals and Alloys*, Butterworths, London, 1962, pp. 35-45.
[12] Evans, U. R., *The Corrosion and Oxidation of Metals: Scientific Principles and Practical Applications*, Edward Arnold, London, 1960, pp. 819-859.
[13] Seward, T. M., *American Journal of Science*, Vol. 274, Feb. 1974, pp. 190-192.
[14] Wilson, P. T., Chemistry Division, Department of Scientific and Industrial Research, Wellington, New Zealand, 1978, personal communication.
[15] Briers, J. D., "Interpretation of Oxidation Colours on Titanium and Other Metals," Report No. 596, Physics and Engineering Laboratory, Department of Scientific and Industrial Research, Petone, New Zealand, Feb. 1978.
[16] Ellis, A. J., *New Zealand Journal of Science*, Vol. 5, No. 4, Dec. 1962, pp. 434-452.

S. D. Cramer,[1] and J. P. Carter[2]

Corrosion in Geothermal Brines of the Salton Sea Known Geothermal Resource Area

REFERENCE: Cramer, S. D. and Carter, J. P., **"Corrosion in Geothermal Brines of the Salton Sea Known Geothermal Resource Area,"** *Geothermal Scaling and Corrosion, ASTM STP 717,* L. A. Casper and T. R. Pinchback, Eds., American Society for Testing and Materials, 1980, pp. 113–141.

ABSTRACT: Corrosion research is being conducted by the Bureau of Mines, U.S. Department of the Interior, to determine suitable construction materials for geothermal resource recovery plants. High chromium-molybdenum iron-base alloys, nickel-base and titanium-base alloys, and a titanium-zirconium-molybdenum alloy (TZM) exhibited good resistance to general, crevice, pitting, and weld corrosion and stress corrosion cracking in laboratory tests in deaerated brines of the Salton Sea known geothermal resource area (KGRA) type at 232°C and in brine containing dissolved carbon dioxide and methane. Only titanium-base alloys were resistant to corrosion in oxygenated Salton Sea KGRA-type brine. Copper adversely affected the resistance to general corrosion of low-alloy steels in deaerated brine, whereas chromium, nickel, silicon, and titanium improved it. Carbon steel, Type 4130 steel, and Types 410 and 430 stainless steels exhibited poor corrosion resistance in field tests in five brine and steam process streams produced from geothermal well Magmamax No. 1. These alloys were highly susceptible to pitting and crevice corrosion. General corrosion rates were high for the carbon and Type 4130 steels.

KEY WORDS: corrosion, crevice corrosion, pitting, stress corrosion cracking, weld corrosion, scaling, geothermal brine, alloys, evaluation

The Imperial Valley of California is one of the major liquid-dominated geothermal regions in the United States. Within this valley are several large and distinct known geothermal resource areas (KGRAs) that contain substantial quantities of potentially recoverable minerals, metals, and energy. One such area, the Salton Sea KGRA, contains brines that have a

[1] Chemical engineer, Avondale Research Center, Bureau of Mines, U.S. Department of the Interior, Avondale, Md. 20782.
[2] Research chemist, Avondale Research Center, Bureau of Mines, U.S. Department of the Interior, Avondale, Md. 20782.

high mineral content, up to 30 weight percent total dissolved solids [1],[3] and temperatures up to 350°C [2]. These brines are among the most corrosive naturally occurring fluids, and the recovery of resources from them will severely challenge existing materials technology.

In 1974, the Bureau of Mines, of the U.S. Department of the Interior, undertook an intensive program to identify construction materials for geothermal resource recovery plants. Field research and testing were conducted to determine the corrosion resistance of commercially available metals and alloys [3,4], the chemistry of flowing geothermal brines [5], suitable electrochemical techniques for studying corrosion in scale-forming brines [6], scale deposition kinetics [7], and fluid sampling techniques for two-phase geothermal fluids [8]. Laboratory research was conducted to study further the corrosion resistance of commercially available metals and alloys [9-11] and to determine the solubility of noncondensable gases in geothermal brines [12,13]. By leading to an improved selection of materials, reduced maintenance costs, and extended equipment service life, such research will help conserve valuable mineral and metal resources and support the development of new technology for the recovery of mineral, metal, and energy values from geothermal brines.

This report describes the results of laboratory and field corrosion tests in brines typical of the Salton Sea KGRA. Laboratory results for 41 alloys are characterized in terms of general, pitting, and crevice corrosion, stress corrosion cracking, and weld corrosion. The effects of dissolved gases and pH on a number of these alloys is discussed. Field results for Type 1020 carbon steel, Type 4130 steel, and Types 410 and 430 stainless steels, from a recent test of over 20 alloys in five distinct geothermal environments, are characterized in terms of general, pitting, and crevice corrosion.

Experimental Procedure

Laboratory Tests

The laboratory testing procedure has been described thoroughly elsewhere [10,11]. The tests were conducted in a nonrecirculating, 2-litre, glass-lined autoclave at a temperature of 232 ± 2°C. The autoclave was two-thirds-filled with a synthetic Salton Sea KGRA-type geothermal brine (Table 1), with an initial pH of 6.1. The autoclave and brine were initially deaerated at room temperature by evacuation at the vapor pressure of the brine for 3 min. The residual oxygen present in the brine following this treatment was estimated to be less than 0.1 ppm. In tests involving dissolved gases, the brine was heated to 232°C and pure gas was injected into the autoclave at an overpressure determined from gas solubility data to give a

---

[3] The italic numbers in brackets refer to the list of references appended to this paper.

TABLE 1—*Composition of geothermal brines.* [a]

Constituent	Salton Sea KGRA-type Brine (pH 6.1), ppm	Magmamax No. 1 Wellhead Brine, Sept.-Nov. 1976 (pH 5.1), ppm
Na	53 000	47 900
Ca	28 800	22 100
K	16 500	8 800
Fe	2 000	265
Mn	1 370	620
Zn	500	290
Sr	440	725
Si	190	240
B	390	[c]
Ba	250	140
Li	210	170
Pb	80	50
Rb	70	[c]
Cs	20	[c]
Mg	10	95
Cu	[b]	0.5
Ag	[b]	0.6
HCO$_3$	[b]	800
Cl	155 000	129 000
S	30	[c]

[a] pH measured at 25°C.
[b] 3 ppm copper, 500 ppm CO_2, and <1 ppm silver were reported in the original analysis [1].
[c] Not determined.

known concentration of gas in the brine. These concentrations were: carbon dioxide, 250 ppm; methane, 100 ppm; and oxygen, 100 ppm. The duration of the laboratory tests was 15 and 30 days, except for tests involving dissolved oxygen, where failure of the Type 316 stainless steel autoclave components shortened exposures to 5 to 7 days.

Corrosion specimens were cut from sheets measuring up to 0.3 cm thick. The specimens for general corrosion measurements (weight loss) were 2.5 by 2.5-cm coupons. The crevice corrosion specimens were two 5 by 1.3-cm overlapped coupons bound together by Teflon[4] straps and separated by a 50-μm-wide gap. The stress corrosion specimens were 5 by 0.6-cm U-bends that had been bent around a mandrel with a 1.3-cm radius of curvature. The legs of the stress corrosion specimens were then compressed further to fit into a slotted mullite holder. The weld specimens were 2.5 by 2.5-cm weight-loss coupons sheared from two 1.3-cm-wide strips that had been butt-welded together, using the shielded metal arc (SMAW) or tungsten

[4] The use of trade names is for identification purposes only and does not imply endorsement by the Bureau of Mines.

inert-gas arc (GTAW) welding technique. Various welding rods (SMAW) or filler alloys (GTAW) were used to weld the specimens.

The general, crevice, and stress corrosion specimens were lightly surface ground and edge ground to a 120-grit finish. The weld specimens were sand blasted after being welded, and the edges were ground to a 120-grit finish. Select weld specimens were heat treated in flowing argon for 1 h at the following temperatures to relieve residual stresses: Type 1020 carbon steel, 625°C; Type 4130 steel, 525°C; Type 316 L stainless steel and E-Brite 26-1, 1100°C (full anneal). The carbon steel and Type 4130 steel specimens were air cooled, while Type 316 L stainless steel and E-Brite 26-1 were water quenched. The Type 316 L stainless steel and E-Brite 26-1 weld specimens were not exposed to organic solvents prior to welding or heat treatment, nor were the E-Brite 26-1 weld specimens exposed to nitrogen during welding and heat treatment. Final preparation of the specimens involved cleaning with pumice, washing with demineralized water, rinsing with methanol, drying, and weighing.

During the tests, the specimens were suspended in the brine from glass holders. Duplicate specimens were included in each test. General, crevice, and stress corrosion specimens for a given alloy were usually included in the same test. Nonweld Type 1020 carbon steel general corrosion specimens were included as controls in tests on welded specimens and in all tests on low-alloy steels. At the conclusion of the tests, the specimens were mechanically cleaned with a hard rubber stopper to remove corrosion products. Light chemical cleaning was further required for certain alloys. The specimens were then washed with demineralized water, rinsed with methanol, dried, and weighed.

Field Tests

Field corrosion tests were conducted at the Bureau's Geothermal Test Facility (GTF), located on the Salton Sea KGRA near Calipatria, Calif. Wellhead brine from geothermal well Magmamax No. 1 (Table 1) was used to produce brine and steam process streams characteristic of those in geothermal resource recovery plants. Corrosion tests were conducted in wellhead brine (P1), in the steam (P3) and concentrated brine (P2) following a first-stage separation of steam from the wellhead brine, and in steam (P5) and concentrated brine (P4) following a second-stage separation of steam from the first-stage concentrated brine (P2). The first-stage and second-stage steam separators were operated at successively lower pressures, so that significant fractionation of the brine fed stream was achieved in the separators. The operation of the GTF during an earlier test is discussed by Carter and coworkers [4].

The field corrosion tests were conducted for 15, 30, and 45 days' duration. The mean conditions for the five process streams for 45 days of operation are reported in Table 2. The average flow rate of brine into the GTF for this period was 130 litres/min. Although gas concentrations in the steam process streams were not analyzed, it is assumed that much of the carbon dioxide and methane, and perhaps hydrogen sulfide and ammonia, present in the brine was removed from the brine in the first-stage steam separator.

The corrosion specimens were 4.4 by 2.5-cm weight-loss coupons sheared from sheets 0.3 cm thick or less. They were edge ground to a 120-grit finish. Treatment of the carbon steel and low-alloy steel specimens also included chemical descaling at 60°C in 12 percent (by volume) sulfuric acid containing 2.5 ml of Rodine 95 inhibitor per litre. Following this, the specimens were washed with demineralized water, rinsed with methanol, dried, and weighed. The specimens were mounted, as in the earlier field test [4], on 2.4-m-long metal bars, using titanium dioxide (TiO_2) insulators and Teflon washers to prevent contact with the bar and galvanic corrosion. The bars were axially loaded into 7.6-cm diameter horizontal pipe sections through which the process streams flowed. At regular intervals, bars were removed and replaced with new bars holding fresh specimens until the scheduled tests of 15, 30, and 45 days' duration were completed. In this way, more than 4000 specimens of over 20 alloys were tested in the five process environments. Results for the 16 iron, titanium, nickel, and copper-base alloys from the field tests not covered here will be reported later as they become available.

At the conclusion of the tests, the specimens were covered with scale consisting of minerals deposited from the process stream and corrosion product. The composition of this scale depended upon the process stream in which it was produced, as in the earlier field test [4]. The scale was removed by lightly scraping the specimen, followed by chemical cleaning. Two procedures were used in chemically cleaning the four alloys reported here. The Type 1020 carbon steel and Type 4130 steel specimens were cleaned in the inhibited sulfuric acid solution. The Types 410 and 430

TABLE 2—*Mean operating conditions for corrosion test packages in Salton Sea KGRA field test.* [a]

Corrosion Test Package	Temperature, deg C	Pressure, psia[b]	Chloride Concentration, ppm	pH[c]
Wellhead brine (P1)	215	291	115 000	5.3
Brine from 1st separator (P2)	199	237	127 000	5.7
Steam from 1st separator (P3)	199	237	8 000	6.2
Brine from 2d separator (P4)	180	148	129 000	5.8
Steam from 2d separator (P5)	180	148	1 700	6.9

[a] Wellhead brine flow rate = 130 litres per min.
[b] Multiply by 6.89×10^{-3} to convert psi to MPa.
[c] Measured at 25°C.

stainless steel specimens were cleaned at 60°C in 20 percent (by volume) nitric acid containing, when necessary to facilitate scale removal, a few drops of hydrofluoric acid.

Specimen Evaluation

General corrosion rates were computed based on the total surface area of the specimens. In the laboratory tests, crevice corrosion specimens were examined with a microscope to define the area of attack. This area was used to compute the corrosion rate for the crevice, and crevice corrosion was determined by comparing the corrosion rate in the crevice with the general corrosion rate. In the field tests, crevice corrosion was determined by microscope examination of the specimen in the area near the crevice formed by the Teflon mounting washers. Evidence of preferential corrosion in this area indicated susceptibility to crevice corrosion. Pits with depths greater than 12 μm were measured with an optical micrometer. The maximum pitting rates were based on the deepest pit, and the average pitting rates were based on the mean depth of the five deepest pits (laboratory tests) and ten deepest pits (field tests). The U-bend laboratory specimens were examined with a microscope for cracks, and, if none were detected, the legs of the specimen were closed, and the specimen was reexamined. Stress corrosion cracking was reported when cracks were observed in either case.

Results

Laboratory Tests

General Corrosion—General corrosion rates are given in Table 3 for 15-day exposures of nonweld specimens in deaerated Salton Sea KGRA-type brine and brine containing dissolved methane, carbon dioxide, or oxygen. Select tests of 30 days' duration gave results similar to those for 15-days' exposure and are not reported. The result for Type 1020 carbon steel in deaerated brine is the average for 50 specimens and has a standard deviation of ±200 μm/year. The results for duplicate specimens of carbon steel in the same test agreed well, but small differences in the test conditions, for example, in deaeration and temperature, apparently led to substantial differences in the measured corrosion rates for different tests. With the exceptions of Type 1020 carbon steel, Mariner steel, Type 4130 steel, Monel 400, and 70/30 cupronickel, the general corrosion rates for the alloys in deaerated brine were less than 125 μm/year.

The addition of methane and carbon dioxide to the brine had a relatively small effect on general corrosion rates. Methane tended to reduce corrosion rates, whereas carbon dioxide increased the corrosion rates for the iron-

TABLE 3—*General corrosion rates in Salton sea KGRA-type brine at 232°C, μm/year.* [a]

Alloy	Deaerated	CH_4, 100 ppm	CO_2, 250 ppm	O_2, 100 ppm
Iron base				
1020 carbon steel	419	46	[b]	26 900
Cor-Ten B	[b]	10	46	30 000
Mariner	140	38	210	33 300
4130 steel	330	84	[b]	25 400
430 stainless	0	0	[b]	8 760
316 L stainless	0	0	5	4 010
E-Brite 26-1	3	0	5	530
Carpenter 20 stainless	8	[b]	[b]	[b]
Sandvik 3RE60	53	[b]	[b]	[b]
Nickel base				
Monel 400	500	0	3	7 320
Inconel X-750	8	[b]	[b]	[b]
Inconel 625	0	3	3	490
Hastelloy S	0	[b]	[b]	[b]
Hastelloy G	3	[b]	[b]	[b]
Hastelloy C-276	0	0	[b]	150
Titanium base				
Ti-50A	5	0	0	0
Ti-1.7W	0	[b]	[b]	[b]
Ti-2Ni	3	[b]	[b]	0
Ti-10V	0	[b]	[b]	[b]
Ti-0.2Pd	0	0	0	0
TiCode-12	0	0	[b]	0
Copper base				
70/30 cupronickel	240	30	25	5 720
Molybdenum base				
TZM	15	[b]	[b]	[b]

[a] Divide by 25.4 to convert from μm per year to mils per year.
[b] Not tested.

base alloys. Intergranular corrosion was observed on Mariner steel exposed to brine containing carbon dioxide. The addition of oxygen (100 ppm) sharply increased the general corrosion rates for the iron, nickel, and copper-base alloys. Except for carbon steel and the low-alloy steels, in which corrosion was severe but relatively uniform, the increased corrosion rates were largely caused by nonuniform attack; for example, substantial areas of the specimens were deeply corroded while the remaining areas appeared protected or were less severely attacked. The titanium alloys were highly resistant to general corrosion in brine containing dissolved oxygen.

A comparative study of the general corrosion of low-alloy steels in deaerated brine was conducted using Type 1020 carbon steel as a control. The concentrations of the principal alloying elements in the low-alloy steels—chromium, molybdenum, titanium, silicon, nickel, and copper—are shown in Table 4. The experimental corrosion rates were normalized to correct for the small differences in test conditions noted earlier by a simple ratio with the rate for the corresponding Type 1020 carbon steel controls (Table 5). The normalization was based on a corrosion rate of 419 μm/year for the carbon steel. After normalization, the rates were fit by the least-squares method to an empirical model involving the composition of the alloys. The fitted corrosion rates are given in Table 5 and will be discussed later.

The effect of pH on the general corrosion rates of carbon and low-alloy steels is shown in Table 6. The reported pH is the value measured at the beginning of each test at 25°C. Regardless of the initial pH, the pH of the deaerated brine at the end of the 15-day tests ranged between 5.1 and 5.7. (For oxygenated brine with an initial pH of 6.1, the final pH ranged between 2.8 and 4.3.) The corrosion rates did not change significantly between pH 6.1 and pH 3.0. However, they rose rapidly between pH 3.0 and 2.0. At most pH values, the corrosion rates for the low-alloy steels were appreciably lower than those for carbon steel. In oxygenated brine at pH 1.7, the corrosion rates for Cor-Ten B and Mariner steel were 65 000 and 53 200 μm/year, respectively, roughly double the value at pH 6.1 (Table 3).

Crevice Corrosion—Crevice corrosion results are given in Table 7 for 15 and 30-day exposures of nonweld specimens in deaerated Salton Sea KGRA-type brine and in brine containing dissolved methane, carbon dioxide, or oxygen. Crevice corrosion of Ranks 2 and 3 was observed in 40 percent of the alloys tested in deaerated brine. Crevice corrosion was not

TABLE 4—*Principal alloying elements in low-alloy steels (in weight percent).*

Alloy	Cr	Mo	Ti	Si	Ni	Cu
Iron	0	0	0	0	0	0
1020 carbon steel	0	0	0	0	0	0
Fe-0.05 Mo	0	0.05	0	0	0	0
Fe-1Mo	0	0.85	0	0	0	0
Fe-2Mo	0	1.36	0	0	0	0
Fe-5Mo	0	3.56	0	0	0	0
Fe-10Mo	0	10.00	0	0	0	0
Fe-2.25Cr-1Mo	2.11	0.91	0	0	0	0
Fe-2.25Cr-2Mo	2.44	1.71	0	0	0	0
Fe-2.25Cr-2Mo-1Ti	2.40	1.95	1.38	0	0	0
Fe-2.25Cr-1Mo-3Ti	2.02	0.92	2.59	0	0	0
Fe-2.25Cr-1Mo-1Si	2.14	0.80	0	0.88	0	0
Fe-2.25Cr-1Mo-1Si-1Ti	2.13	0.72	1.11	0.84	0	0
Croloy ½	0.67	0.56	0	0.15	0.11	0.08
Croloy 1¼	1.26	0.54	0	0.74	0.23	0.12
Croloy 3	2.96	0.94	0	0.29	0.21	0.16
Croloy 5	4.52	0.55	0	0.38	0.20	0.11
Croloy 7	6.61	0.53	0	0.67	0.23	0.10
Croloy 9	7.97	1.13	0	0.28	0.20	0.07
4130 steel	0.95	0.12	0	0.30	0	0

TABLE 5—*General corrosion of low-alloy steels in deaerated Salton Sea KGRA-type brine at 232°C.*

Alloy	Corrosion Rate, μm/year Normalized[a]	Fitted	Deviation, percent
Iron	401	531	−24
1020 carbon steel	419	531	−21
Fe-0.05Mo	564	531	6
Fe-1Mo	551	533	3
Fe-2Mo	579	533	8
Fe-5Mo	747	538	38
Fe-10Mo	480	556	−14
Fe-2.25Cr-1Mo	185	229	−19
Fe-2.25Cr-2Mo	262	203	29
Fe-2.25Cr-2Mo-1Ti	61	119	−48
Fe-2.25Cr-1Mo-3Ti	86	71	21
Fe-2.25Cr-1Mo-1Si	89	124	−28
Fe-2.25Cr-1Mo-1Si-1Ti	96	61	57
Croloy ½	658	577	14
Croloy 1¼	257	302	−15
Croloy 3	488	533	−8
Croloy 5	231	196	18
Croloy 7	203	155	31
Croloy 9	274	312	−12
4130 steel	333	333	0

[a] Data normalized using a corrosion rate for carbon steel of 419 μm/year.

observed in Inconel 625 in 15-day exposures but appeared in 30-day exposures. Titanium 50A exhibited severe crevice corrosion in the deaerated brine. No crevice corrosion was observed in Alloys Fe-1Mo, Fe-2Mo, Fe-2.25Cr-1Mo, and Fe-2.25Cr-2Mo. Severe crevice corrosion was observed in the carbon and low-alloy steels in brines containing methane. In brines containing carbon dioxide, severe crevice corrosion was observed only in Mariner steel. Susceptibility to crevice corrosion increased sharply for Type 316 L stainless steel, E-Brite 26-1, Inconel 625, Hastelloy C-276, and 70/30 cupronickel in brines containing oxygen.

The effect of pH on the crevice corrosion of carbon and low-alloy steels in deaerated brine is shown in Table 6. Susceptibility to crevice corrosion increased to severe for Type 1020 carbon steel at pH 4 and below. Mariner steel exhibited severe crevice corrosion only at pH 1. Crevice corrosion of Type 4130 steel was not severe.

Pitting Corrosion—Pitting corrosion rates are given in Table 8 for 15-day exposures in deaerated Salton Sea KGRA-type brine and brine containing dissolved methane, carbon dioxide, or oxygen. Pitting corrosion was observed in half the alloys tested in deaerated brine. The low-alloy steels were particularly susceptible to pitting, as were several titanium-base alloys. With

TABLE 6—*Effect of pH on the corrosion of mild and low-alloy steels in deaerated Salton Sea KGRA-type brine at 232°C.*

pH	Type 1020 Carbon Steel	Type 4130 Steel	Cor-Ten B	Mariner
\multicolumn{5}{c}{General Corrosion Rate, μm/year[a]}
6.1	419	333	[b]	140
5.0	267	140	132	279
4.0	445	193	157	254
3.0	401	183	200	272
2.0	1240	566	419	488
1.0	6450	4670	6860	3780
\multicolumn{5}{c}{Crevice Corrosion[c]}
6.1	0	[b]	3	0
5.0	0	0	[b]	[b]
4.0	3	1	[b]	1
3.0	[b]	0	[b]	0
2.0	3	2	[b]	1
1.0	3	1	[b]	3
\multicolumn{5}{c}{Pitting Corrosion[d]}
6.1	1680/970	[b]	1090/[b]	[b]
5.0	2110/790	790/580	530/460	[b]
4.0	[b]	[b]	1120/790	890/710
3.0	[b]/410	660/530	740/580	[b]
2.0	1140/860	[b]	[b]	[b]
1.0	2670/1730	3940/2790	4450/3430	3180/2510

[a] General corrosion rates normalized to 419 μm/year for Type 1020 carbon steel at pH 6.1.
[b] Not measured.
[c] Number scale for rating crevice corrosion results:
  0 = none.
  1 = penetration < 25 μm/year.
  2 = 25 μm/year < penetration < 130 μm/year.
  3 = penetration > 130 μm/year.
[d] Pitting results reported as the maximum pitting rate followed by the average pitting rate (in μm/year).

the exception of Monel 400, the nickel-base alloys were generally resistant to pitting. Although the average pitting rates in deaerated brine were similar for iron-molybdenum and iron-chromium-molybdenum alloys, the maximum pitting rates were substantially higher for the iron-molybdenum alloys.

Pitting rates for Type 1020 carbon steel decreased in brine containing methane. In brine containing carbon dioxide, pitting rates for Type 1020 carbon steel and Mariner steel increased substantially compared with those measured in brine containing methane. The pitting rates for Alloy Ti-50A in brine containing methane were low compared with those in deaerated brine. Two types of pits were formed in E-Brite 26-1 exposed to brine

TABLE 7—*Crevice corrosion in Salton Sea KGRA-type brine at 232°C.* [a]

Alloy	Deaerated	CH$_4$, 100 ppm	CO$_2$, 250 ppm	O$_2$, 100 ppm
Iron base				
1020 carbon steel	0[b]	3	0	0
Cor-Ten B	3	3	0	0
Mariner	0	3	3	0
4130 steel	3	3	c	c
430 stainless	0	0	c	c
316 L stainless	1	0	0	3
E-Brite 26-1	0	0	0	3
Carpenter 20 stainless	2	c	c	c
Sandvik 3RE60	3	c	c	c
Nickel base				
Monel 400	0	0	0	0
Inconel X-750	2	c	c	c
Inconel 625	2	0	0	3
Hastelloy S	1	c	c	c
Hastelloy G	1	c	c	c
Hastelloy C-276	0	1	c	3
Titanium base				
Ti-50A	3	1	0	0
Ti-1.7W	1	c	c	c
Ti-2Ni	2	c	c	0
Ti-10V	1	c	c	c
Ti-0.2Pd	0	0	1	0
TiCode-12	c	0	0	0
Copper base				
70/30 cupronickel	0	0	0	3
Molybdenum base				
TZM	0	c	c	c

[a] Number scale for ranking crevice corrosion results:
 0 = none.
 1 = penetration <25 μm/year.
 2 = 25 μm/year <penetration <130 μm/year.
 3 = penetration > 130 μm/year.
[b] Pits in crevice.
[c] Not measured.

containing methane and carbon dioxide—broad, shallow pits and narrow, deep pits (Fig. 1). The pitting rates for all alloys tested in brine containing oxygen were high except for the titanium-base alloys. E-Brite 26-1 was extremely susceptible to pitting in oxygenated brine. The pitting rates for Cor-Ten B in oxygenated brine at pH 1.7 were 27 900/23 700 μm/year compared with 16 000/8 000 μm/year at pH 6.1.

The effect of pH on pitting of carbon and low-alloy steels in deaerated brine is shown in Table 6. Between pH 6.1 and 2.0, pitting rates for Type 1020 carbon steel were generally higher than those for the low-alloy steels but did not change appreciably with pH. However, maximum and average pitting rates for each steel increased sharply for pH 1.0.

TABLE 8—*Pitting corrosion rates in Salton Sea KGRA-type brine at 232°C.* [a]

Alloy	Deaerated	CH_4, 100 ppm	CO_2, 250 ppm	O_2, 100 ppm
Iron base				
1020 carbon steel	1700/970	1100/660	3300/1700	35 100/13 700
Cor-Ten B	1090/[b]	710/530	530/480	16 000/8 000
Mariner	[b]	530/510	1300/810	2 900/2 500
4130 steel	[b]/1200	580/530	[b]	[b]
Fe-1Mo	2000/1200	[b]	[b]	[b]
Fe-2Mo	2000/1100	[b]	[b]	[b]
Fe-2.25Cr-1Mo	1100/890	[b]	[b]	[b]
Fe-2.25Cr-2Mo	1200/1100	[b]	[b]	[b]
430 stainless	0/0	D	D	D
316 L stainless	0/0	0/0	0/0	5 100/3 300
E-Brite 26-1	0/0	790/[b]	690/660	24 300/21 000
Carpenter 20 stainless	0/0	[b]	[b]	[b]
Sandvik 3RE60	0/0	[b]	[b]	[b]
Nickel base				
Monel 400	2400/1600	2900/1500	D	6 000/3 800
Inconel X-750	0/0	[b]	[b]	[b]
Inconel 625	0/0	0/0	0/0	13 500/[b]
Hastelloy S	0/0	[b]	[b]	[b]
Hastelloy C-276	0/0	0/0	0/0	16 300/[b]
Titanium base				
Ti-50A	5500/1200	480/430	0/0	D
Ti-1.7W	0/0	[b]	[b]	[b]
Ti-2Ni	1100/710	[b]	[b]	0/0
Ti-10V	3000/1800	[b]	[b]	[b]
Ti-0.2Pd	1200/860	1100/860	0/0	0/0
TiCode-12	[b]	0/0	0/0	0/0
Copper base				
70/30 cupronickel	0/0	1200/1100	1200/610	9 500/8300
Molybdenum base				
TZM	0/0	[b]	[b]	[b]

[a] Pitting results are reported as the maximum pitting rate followed by the average pitting rate (in μm/year), or as detected (D) for pitting rates less than 340 μm/year.
[b] Not measured.

Stress Corrosion—Stress corrosion results are given in Table 9 for 15 and 30-day exposures of U-bend specimens in deaerated Salton Sea KGRA-type brine and brine containing methane, carbon dioxide, or oxygen. Stress corrosion cracking in the deaerated brine was detected in the iron-base alloys (Types 430, 316 L, and Carpenter 20 stainless steels, and Sandvik 3RE60), and in several nickel-base alloys (Inconel 625, and Hastelloy G), the latter only in 30-day exposures. Stress corrosion cracking in brine containing methane or carbon dioxide occurred only in Types 430 and 316 L stainless steel and Hastelloy C-276. Alloy E-Brite 26-1 did not exhibit stress corrosion cracking in deaerated brine or brine containing methane, carbon dioxide, or oxygen.

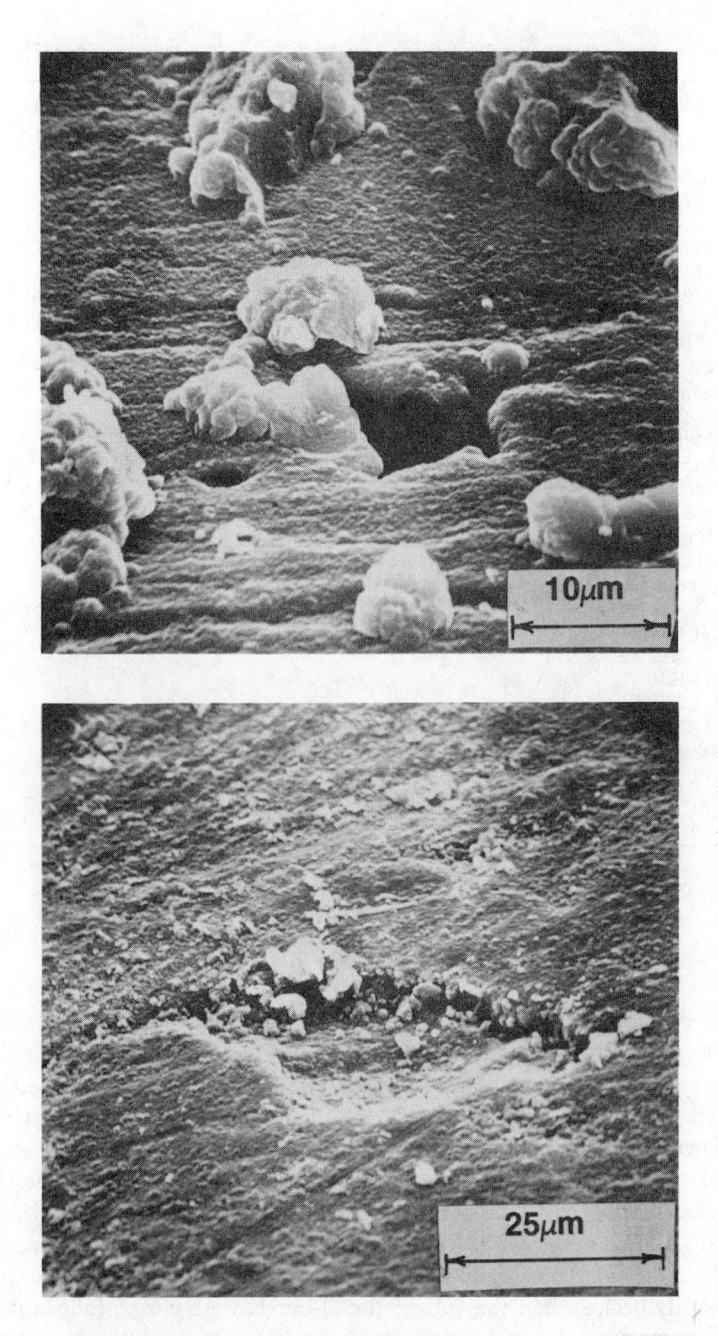

FIG. 1—*Pitting corrosion of E-Brite 26-1 at 232°C in Salton Sea KGRA-type brine containing dissolved methane.*

TABLE 9—*Stress corrosion cracking (SCC) in Salton Sea KGRA-type brine at 232°C.*[a]

Alloy	Deaerated	CH_4, 100 ppm	CO_2, 250 ppm	O_2, 100 ppm
Iron base				
1020 carbon steel	...	...	b	b
Cor-Ten B	...	...	...	...
Mariner	b	b	...	b
4130 steel	...	b	b	b
430 stainless	SCC	SCC	SCC	SCC
316 L stainless	SCC	SCC	SCC	SCC
E-Brite 26-1	...	...	...	...
Carpenter 20 stainless	SCC	b	b	b
Sandvik 3RE60	SCC	b	b	b
Nickel base				
Monel 400	...	...	b	b
Inconel X-750	...	b	b	b
Inconel 625	SCC[c]	...	...	SCC
Hastelloy S	...	b	b	b
Hastelloy G	SCC[c]	b	b	b
Hastelloy C-276	...	SCC	...	SCC
Titanium base				
Ti-50A	...	...	...	SCC
Ti-1.7W	...	b	b	b
Ti-2Ni	...	b	b	...
Ti-10V	...	b	b	b
Ti-0.2Pd	...	...	...	...
TiCode-12	b	...	b	b
Copper base				
70/30 cupronickel	...	b	b	b
Molybdenum base				
TZM	...	b	b	b

[a] Dots (...) indicate SCC was not detected.
[b] Not tested.
[c] Thirty-day exposure.

Stress corrosion cracking of Type 316 L stainless steel in oxygenated brine was particularly severe. The cracking was transgranular (Fig. 2). (The etched grain boundaries shown in Fig. 2 were present in the as-received Type 316 L stainless steel and were not due to corrosion in the brines.) Stress corrosion cracking of Hastelloy C-276 was initiated at pits in the bends of the specimens. The cracking appeared to be transgranular (Fig. 3). Cracks in Inconel 625 also appeared to be transgranular (Fig. 4). The cracks in Alloy Ti-50A were located at small areas of localized corrosion along the edges of the U-bend specimens (Fig. 5). The Alloy Ti-50A specimens readily broke when the legs of the U-bends were closed following the test. Examination of the fracture surfaces showed that the cracks had propagated from these small areas of localized corrosion. However, since the Alloy Ti-50A specimens had been rinsed with methanol prior to the

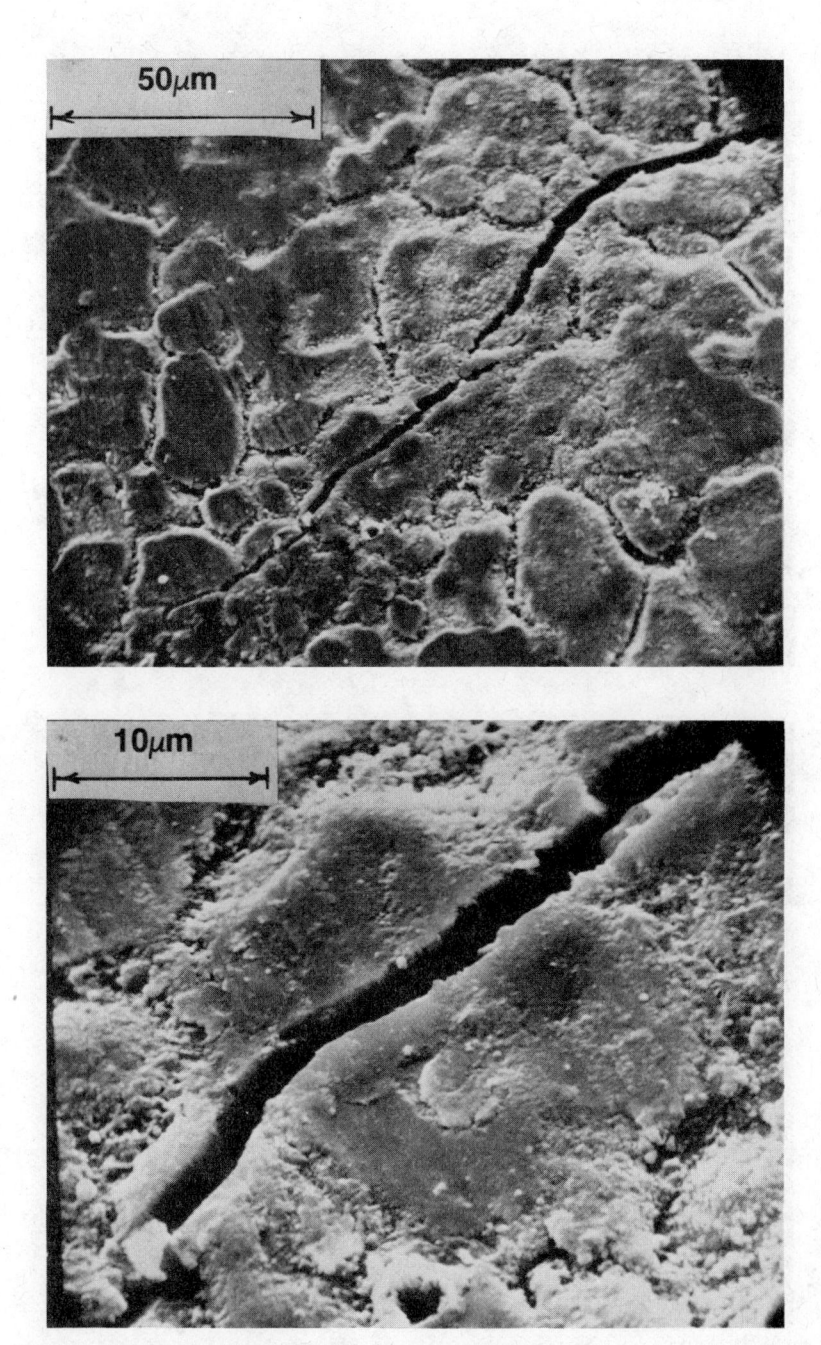

FIG. 2—*Transgranular stress corrosion cracking of Type 316 L stainless steel in oxygenated Salton Sea KGRA-type brine at 232°C.*

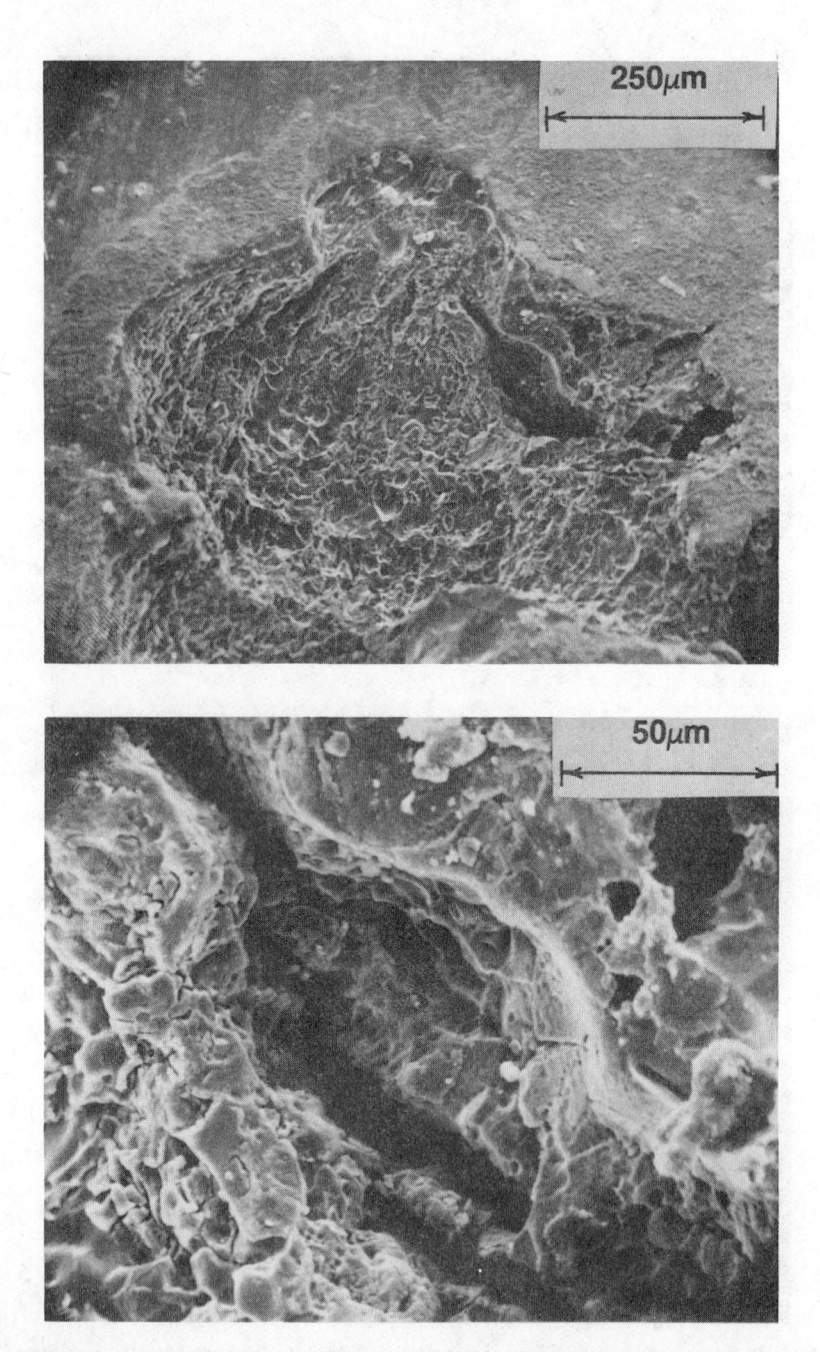

FIG. 3—*Stress corrosion crack in Hastelloy C-276 in oxygenated Salton Sea KGRA-type brine at 232°C.*

FIG. 4—*Stress corrosion crack in Inconel 625 in oxygenated Salton Sea KGRA-type brine at 232°C.*

test and since titanium exhibits stress corrosion cracking in methanol [14], the exact cause of the cracking in the oxygenated brine was not clear.

Weld Corrosion—The general corrosion rates are given in Table 10 for 15-day exposures of weld specimens in deaerated Salton Sea KGRA-type brine and brine containing oxygen. The general corrosion rates were based on the nominal surface area of the 2.5 by 2.5-cm weight-loss specimens. Both the specimen alloy and the alloy used to fill the butt weld are listed in Table 10. Except for carbon steel and Type 4130 steel, the filler alloys were identical to the specimen alloys, and the welds were made using the GTAW technique. Carbon steel and Type 4130 steel were welded using several types of welding rods by the SMAW technique.

The results for Type 1020 carbon steel in deaerated brine suggest that the weld corrodes at a rate similar to the nonweld specimens. However, the weld area (including the adjacent heat-affected zone), where most of the increased corrosion occurred, corresponds to roughly one fifth of the 2.5 by 2.5-cm weld specimen, and an increase in the measured corrosion rate of 25 μm/year represents an increase in corrosion rate for the weld area of roughly 125 μm/year. Using this as a basis, the corrosion rate of welds

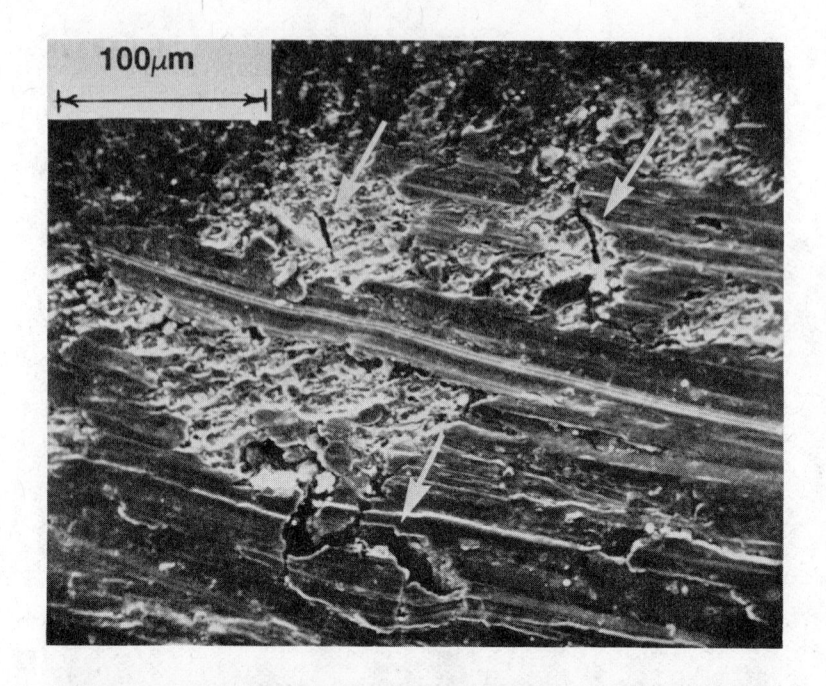

FIG. 5—*Stress corrosion crack in titanium 50A exposed to oxygenated Salton Sea KGRA-type brine at 232°C.*

TABLE 10—*General corrosion of weld and nonweld specimens in Salton Sea KGRA-type brine at 232°C, μm/year.*

Alloy/Filler Alloy	Welding Technique	Nonweld[a]	Weld	Weld, Heat Treated
		Deaerated		
1020 carbon steel/E6010	SMAW	419	510	520
1020 carbon steel/E6011	SMAW	419	380	[b]
1020 carbon steel/E6012	SMAW	419	420	430
1020 carbon steel/E6013	SMAW	419	510	540
1020 carbon steel/E7016	SMAW	419	520	410
1020 carbon steel/E7018	SMAW	419	430	460
1020 carbon steel/E7024	SMAW	419	600	420
4130 steel/E6010	SMAW	330	430	470
Ti-50A/Ti-50A	GTAW	5	70	[b]
Ti-0.2Pd/Ti-0.2Pd	GTAW	0	10	[b]
TiCode-12/TiCode-12	GTAW	0	10	[b]
		Oxygen, 100 ppm		
1020 carbon steel/E6010	SMAW	26 900	30 000	[b]
4130 steel/E6010	SMAW	25 400	28 400	[b]
316 L stainless/ 316 L stainless	GTAW	4 010	3 700	1 200
E-Brite 26-1/E-Brite 26-1	GTAW	530	3 700	2 100
Ti-50A/Ti-50A	GTAW	0	0	[b]
Ti-0.2Pd/Ti-0.2Pd	GTAW	0	10	[b]

[a] Nonweld data included for comparison.
[b] Not tested.

made with the E7016 rod would be 920 μm/year. Low-weld corrosion rates, that is, rates comparable to the nonweld specimens, were observed for welds made with E6011, E6012, and E7018 rods. Heat-treating welds made with E7016 and E7024 rods resulted in low-weld corrosion rates. Welds in Type 4130 steel made with the E6010 rod corroded at rates substantially higher than the nonweld specimens.

In oxygenated brine, the difference between the corrosion rate of the weld and nonweld carbon steel specimens was not significant because of the high general corrosion rates. This was also true of the Type 4130 steel specimens. There was no increase in the corrosion of Ti-50A and Ti-0.2Pd welds. Welding did not increase the corrosion rate of Type 316 L stainless steel, but heat treating had the overall effect of reducing the general corrosion rate of the alloy. Stress corrosion cracks were present across the weld in specimens of Type 316 L stainless steel that had not been heat treated. Welding substantially increased the corrosion rate of E-Brite 26-1 and heat treatment was only partially effective in lowering the rate.

Field Tests

General Corrosion—Table 11 reports general corrosion rates for four ferrous alloys in 15, 30, and 45-day tests in brine and steam process streams, using wellhead brine from geothermal well Magmamax No. 1. In all five process streams, the severity of attack, in decreasing order, was Type 1020 carbon steel, Type 4130 steel, and Types 410 and 430 stainless steels. Wellhead brine was the most corrosive process stream. In general, the corrosion rates decreased with increasing exposure time, which reflected the buildup of scale and corrosion product on the surface of the specimens. However, the 45-day corrosion rates for Type 1020 carbon steel were still high. General corrosion rates for Type 430 stainless steel were low in each of the process streams except the wellhead brine.

Crevice Corrosion—Crevice corrosion was observed on the four ferrous alloys in all five process streams. The specimens shown in Fig. 6 were exposed to wellhead brine for 45 days. The ring surrounding the mounting holes was partially covered by a Teflon washer that formed a small natural crevice at its perimeter. Much of the corrosion in Types 410 and 430

TABLE 11—*General corrosion rates in Salton Sea KGRA field test, μm/year.*

Alloy	15 Days[a]	30 Days	45 Days
Wellhead brine (P1)			
1020 carbon steel	2210	2510	2410
4130 steel	1170	1470	737
410 stainless	660	787	279
430 stainless	305	175	191
Separated brine (P2)			
1020 carbon steel	1170	914	711
4130 steel	381	356	254
410 stainless	196	213	140
430 stainless	25	112	46
Separated steam (P3)			
1020 carbon steel	1400	1700	991
4130 steel	813	1020	457
410 stainless	279	254	157
430 stainless	89	114	71
Separated brine (P4)			
1020 carbon steel	1910	1470	1020
4130 steel	584	483	119
410 stainless	142	46	170
430 stainless	13	10	5.1
Separated steam (P5)			
1020 carbon steel	432	1090	483
4130 steel	279	457	330
410 stainless	145	117	38
430 stainless	10	23	5.1

[a] Average of four 15-day tests.

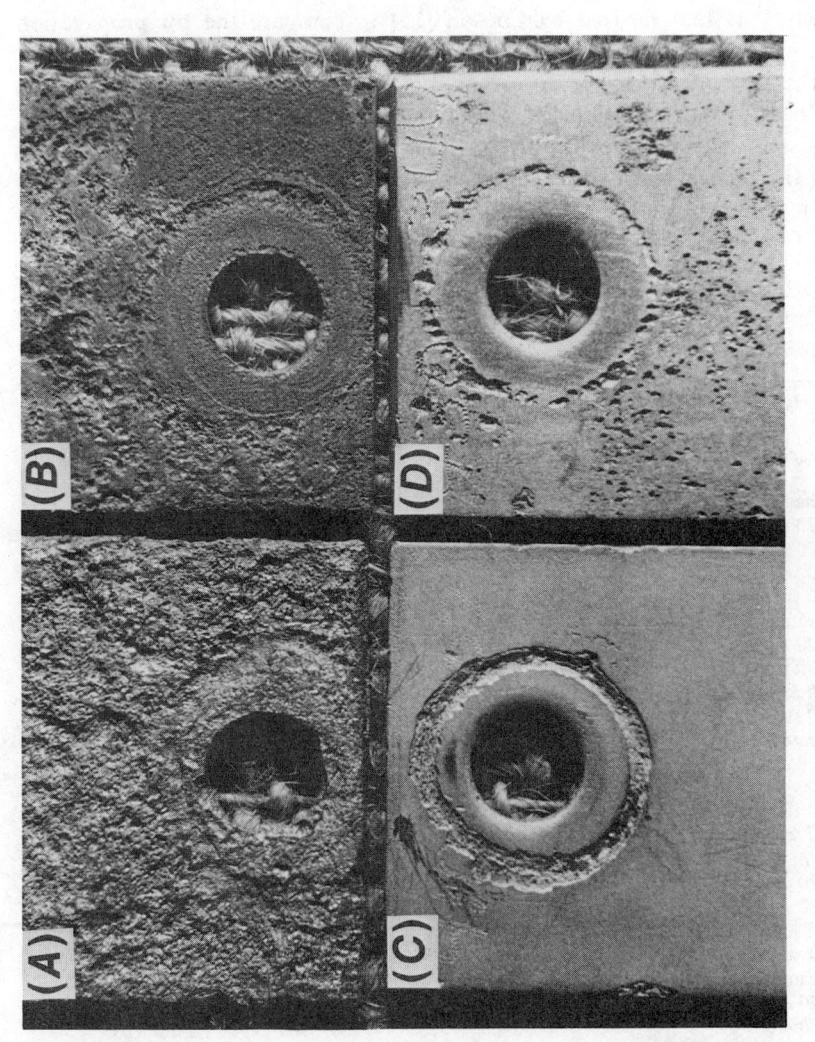

FIG. 6—Specimens of Type 1020 carbon steel (A), Type 4130 steel (B), Type 410 stainless steel (C), and Type 430 stainless steel (D), exposed to Magmamax No. 1 wellhead brine for 45 days, showing evidence of crevice corrosion and pitting.

stainless steels was focused in this area and, by comparison, general corrosion was mild.

Pitting Corrosion—Pitting rates for the four ferrous alloys in 15, 30, and 45-day tests in brine and steam process streams using wellhead brine from geothermal well Magmamax No. -1 are reported in Table 12. The results are given in terms of the maximum and average pitting rates. The pitting "factor" is that devised by Copson [15] to compare the pit propagation rate with the general corrosion rate.

The rate of pitting corrosion was most severe for Type 4130 steel in the wellhead brine, and the pits penetrated completely through specimens of this steel. In the other process streams, Type 1020 carbon steel typically had the highest pitting rate and Type 430 stainless steel the lowest. Pitting rates decreased with increasing exposure time. The maximum pitting rates for Types 410 and 430 stainless steels were not greatly different, and, in

TABLE 12—*Pitting corrosion in Salton Sea KGRA field test.*

Alloy	15 Days[a] Rate[b]	Factor[c]	30 Days Rate[b]	Factor[c]	45 Days Rate[b]	Factor[c]
Wellhead brine (P1)						
1020 carbon steel	14 900/7200	6.7	6 900/4600	2.7	5400/4000	2.2
4130 steel	29 700/8900[d]	25	11 500/5400[d]	7.8	3500/2700	4.7
410 stainless	13 100/7600	20	8 100/5900	10	3100/1900	11
430 stainless	13 000/7400	43	5 000/3800	29	2900/2000	15
Separated brine (P2)						
1020 carbon steel	16 100/7600	14	5 200/3500	5.7	3400/2100	4.8
4130 steel	7900/4200	21	3 900/2700	11	1500/1000	5.9
410 stainless	5700/3600	29	3 800/2500	18	1600/1200	11
430 stainless	6100/2900	240	3 400/2400	30	2000/1400	43
Separated steam (P3)						
1020 carbon steel	17 700/7800	13	7 000/4700	4.1	2400/1900	2.4
4130 steel	10 600/4600	13	11 500/7000[d]	11	2200/1700	4.8
410 stainless	7300/3600	26	4 400/2800	17	2200/1500	14
430 stainless	6100/3800	69	3 400/2500	30	2200/1400	31
Separated brine (P4)						
1020 carbon steel	22 600/9700	12	8 600/6100	5.9	4600/2700	4.5
4130 steel	18 400/5200	32	5 100/2600	11	1400/1100	12
410 stainless	7 000/3200	49	2 100/1400	46	1300/910	7.6
430 stainless	5 200/2200	400	2 600/1900	260	1800/890	350
Separated steam (P5)						
1020 carbon steel	8 300/4700	19	9 200/6900	8.4	1800/1200	3.7
4130 steel	9 100/3400	33	4 200/3100	9.2	1700/1100	5.2
410 stainless	5 000/3000	34	2 400/1700	21	1600/970	42
430 stainless	5 300/1900	530	3 300/2200	140	1100/510	220

[a] Average of four 15-day tests.
[b] Pitting rates reported as the maximum rate followed by the average rate (in μm/year).
[c] Ratio of maximum pitting rate to general corrosion rate.
[d] Pits penetrate through specimens.

fact, for nearly half the results reported in Table 12, the pitting rate of Type 410 stainless steel was less than that of Type 430 stainless steel. Pitting rates in the steam from the first-stage separator (P3) were comparable to those in the brine from the first-stage and second-stage separators (P2 and P4) for each alloy. Pitting rates in the steam from the second-stage separator were lower than those in the steam from the first-stage separator, particularly for Type 1020 carbon steel and Type 4130 steel.

In terms of the pitting factor, the pitting corrosion was most severe in the steam and brine from the second-stage separator. From this viewpoint, Type 430 stainless steel was the alloy most susceptible to pitting, and the specimens were otherwise relatively free of corrosion (except near crevices). The observed weight loss for Type 430 stainless steel was largely due to localized corrosion, and the actual general corrosion rates were probably much lower than those reported in Table 11. Figure 6 illustrates the surface appearance of the alloys exposed to wellhead brine. The Type 1020 carbon steel was rough and heavily corroded, and the low pitting factor was basically a measure of this irregular surface. Type 4130 steel, with a higher pitting factor, clearly showed evidence of pitting, as did Types 410 and 430 stainless steels.

Discussion of Results

General corrosion results from the laboratory (Table 3) and from the field (Table 11 and Refs 3 and 4) differ substantially in the absolute values of the measured corrosion rates. Although Shannon [16] has succeeded with the low-salinity brines in obtaining general corrosion results from the laboratory that are similar to those from the field, reconstituting the high-salinity Salton Sea KGRA brines in the laboratory has been difficult [17]. The laboratory and field tests described here differ in several important ways. For example, geothermal brines from the Salton Sea KGRA contain dissolved hydrogen sulfide and ammonia [18,19]. Ammonia was not present in the laboratory Salton Sea KGRA-type brine, and sulfur, added to the brine as a sulfide, was readily depleted in the nonrecirculating autoclave [10]. The autoclave tests were conducted in a static solution, whereas the field tests were performed in turbulent, rapidly moving flows of brine and steam, which provided high rates of mass transfer to and from the corroding metal surface. The laboratory brine was nonscaling, whereas the brine from Magmamax No. 1 formed scales of varying composition, adherence, coherence, and thickness on the specimens, depending upon the process stream [4]. Even the most dense of these scales had substantial porosity, which sustained contact between the metal and the brine and favored localized corrosion. Two-phase mixtures of brine and steam [8] probably flowed through corrosion test packages P1, P2, and P4, particularly the wellhead brine (P1). Fluctuations in static pressure within the test packages associated with the movement of control valves promoted spontaneous boiling of the flowing brine [6] and accentuated this condition. Laboratory tests represented a single environment, whereas the field tests represented five distinct process environments distinguished by temperature, pressure, fluid phases (steam, brine, two-phase mixture), composition, turbulence, and scaling tendency. Finally, the start-up and shutdown of the field test facility, when specimen bars are changed, can result in oxygen entering the corrosion packages and affecting corrosion, and in cracking and spalling of scales because of cyclic heating and cooling.

Because of these differences, the laboratory results were not expected to reproduce the field results in an absolute sense. Rather they provide data useful for interpreting alloy performance in the field. In doing this, the laboratory tests were focused on the susceptibility of alloys to localized corrosion phenomena—crevice, pitting, and weld corrosion, and stress corrosion cracking—and on comparative studies involving the alloys and the environment (dissolved gases, pH) that do not depend on an absolute measure of corrosion rates but instead require well-controlled test conditions.

The most corrosion-resistant alloys in the laboratory tests using the Salton Sea KGRA-type brine at 232°C and pH 6.1 are listed in Table 13 for (1) deaerated brine and brine containing carbon dioxide and methane and (2) brines containing oxygen. With the exception of Hastelloy C-276, none of the listed alloys exhibited susceptibility to stress corrosion cracking in the respective brines. Carbon steel, Mariner, Cor-Ten B, or Type 4130 steel appear to be unsatisfactory materials for long-term service in the deaerated brine and brine containing carbon dioxide and methane because of high general corrosion rates or susceptibility to pitting and crevice corrosion. Low-alloy steels (such as Fe-2.25Cr-1Mo) containing titanium and silicon have much lower general corrosion rates, may be more resistant to pitting and crevice corrosion, and are in roughly the same cost range as carbon steel, Mariner, Cor-Ten B, and Type 4130 steel.

In deaerated brine and brine containing carbon dioxide and methane, 70/30 cupronickel appeared unsatisfactory because of a high general corrosion rate. Monel 400 also had a high general corrosion rate and was susceptible to pitting corrosion. Carpenter 20, Type 430, and Type 316 L stainless steels, and Sandvik 3RE60, Inconel 625, and Hastelloy G alloys were susceptible to stress corrosion cracking. Carpenter 20 stainless steel, Sandvik 3RE60, and Inconel 625 were also susceptible to crevice corrosion. Inconel X-750 and Alloy Ti-50A were susceptible to crevice corrosion. Alloys Ti-50A and Ti-10V were susceptible to pitting corrosion. In oxygenated brine, the carbon and low-alloy steels, Monel 400 and 70/30 cupronickel, had exceptionally high general corrosion and pitting rates. Alloy E-Brite 26-1 was susceptible to crevice corrosion and had very high pitting rates. Inconel 625, Hastelloy C-276, and Type 316 L stainless steel were sus-

TABLE 13—*Corrosion-resistant alloys in Salton Sea KGRA-type brine at 232°C, pH 6.1 (laboratory).*

Alloy	General Corrosion Rate, μm/year	Crevice Corrosion[a]	Pitting Corrosion Rate,[b] μm/year	Weld Corrosion[c]
\multicolumn{5}{c}{Deaerated Brine and Brine Containing CO_2 and CH_4}
E-Brite 26-1	3	0	0/0	d
Hastelloy S	0	1	0/0	d
Hastelloy C-276[e]	0	0	0/0	d
Ti-1.7W	0	1	0/0	d
Ti-2Ni	3	2	1100/710	d
Ti-0.2Pd	0	0	1200/860	A
TiCode-12	0	d	d	A
TZM	15	0	0/0	d
\multicolumn{5}{c}{Oxygenated Brine}
Ti-2Ni	0	0	0/0	d
Ti-0.2Pd	0	0	0/0	A
TiCode-12	0	0	0/0	d

[a] See Table 7 for code explanation.
[b] See Table 8 for code explanation.
[c] A = corrosion rate of weld is similar to that of nonweld specimen.
[d] Not tested.
[e] Susceptible to SCC in brines containing CH_4.

ceptible to crevice corrosion, pitting corrosion, and stress corrosion cracking. Alloy Ti-50A was susceptible to stress corrosion cracking.

The general corrosion rates in Table 5 for deaerated Salton Sea KGRA-type brine at 232°C and pH 6.1 were fit by least squares to the composition data in Table 4 for carbon and low-alloy steels. The fitted corrosion rates, r (in μm/year), and their deviations from the experimental values are shown in Table 5 for the empirical equation

$$r = 530 - 195[Cr] - 3270[Ni] - 114[Si] - 64[Ti] + 2.6[Mo] + 6780[Cu] + 24[Cr]^2$$

where the elemental concentrations are in weight percent. This approach neglects significant mechanistic and kinetic factors by assuming that the corrosion rate is exclusively a function of alloy composition. Although the fit of this equation is not especially good (standard deviation = 25 percent), it provides an indication of the effect individual alloying elements had on the corrosion rates of the low-alloy steels. For example, the *overall* effect of chromium was to improve the corrosion resistance of the steels. However, because of the quadratic term, which becomes more important at the higher concentrations, the corrosion rate tends to level off with increasing chromium content (up to about 10 percent). Molybdenum increased the corrosion rate slightly, but it has beneficial properties in combination with chromium [20, 21], which reduce susceptibility to pitting and crevice corrosion. Both silicon and titanium were effective in reducing the corrosion rate, although it was not clear whether this would be true in the absence of chromium and molybdenum. Low levels of nickel markedly improved the corrosion resistance of the steels. Low copper levels, on the other hand, substantially decreased the corrosion resistance of the steels. Based on this evidence, copper-bearing steels such as Cor-Ten, Mayari, Croloy, and Mariner should be used cautiously in high-temperature geothermal brines, with the understanding that the copper may adversely affect their performance.

Only Alloy Ti-0.2Pd welds had corrosion rates similar to those of the nonweld specimens in oxygenated brine, Table 13. Carbon steel welded with either E6011, E6012, or E7018 rods gave the best performance in deaerated brine or brine containing carbon dioxide and methane. Heat treatment was not necessary for carbon steel welded with these rods.

Posey and coworkers [22] have examined the corrosion of carbon steel in slowly stirred deaerated 4 M sodium chloride at high temperatures (up to 200°C) from pH 2 to 7. Their results at 200°C were similar to those reported here for carbon and low-alloy steels at 232°C in unstirred deaerated Salton Sea KGRA-type brine: the corrosion rates were relatively unchanged between pH 3 and 7; at lower pH values the corrosion rate increased sharply. Based on their analysis of the kinetics, the rate-limiting step at low pH values (< pH 3) was the diffusion of hydrogen ions to the specimen surface. However, at higher pH values (3 < pH < 6) the iron dissolution and hydrogen evolution reactions were independent of pH. Hence, the corrosion rate was constant and depended only on temperature in this pH range.

The Type 1020 carbon steel, Type 4130 steel, and Types 410 and 430 stainless steels exhibited poor corrosion resistance in the five process streams of the Salton Sea KGRA field test. The general corrosion and pitting rates for Type 1020 carbon steel and Type 4130 steel were high in all five process streams, and the alloys were susceptible to crevice corrosion. The general corrosion rates for Type 410 stainless steel were relatively low in the separated brine and steam from the second separator (P4 and P5), but the pitting rates were high, and the alloy was susceptible to crevice corrosion. Similarly, general corrosion rates for Type 430 stainless steel were low in the five process streams, but pitting rates were high, and the alloy was susceptible to crevice corrosion.

Despite past evidence [9,23] of the severely corrosive conditions in the hypersaline Salton Sea KGRA geothermal brines, some materials used to construct geothermal facilities or recommended for use have been suspect. Goldberg [23] cites the selection for steam separator drums of Type 316 stainless steel cladding, susceptible to pitting and stress corrosion cracking, particularly when oxygen is present as may occur during start-up and shutdown. A more recent example [24] is the recommendation of carbon steel and cast iron for steam separators, direct heat exchangers, and brine pumps in contact with high-salinity brine at 190°C. General corrosion, crevice corrosion, and pitting may be serious problems in such applications. Moreover, the recommendation [24] of copper alloys and, especially, aluminum [9,10], for lower-temperature applications deserves careful examination if materials problems are to be avoided.

Summary

General, crevice, pitting, and weld corrosion and stress corrosion cracking data were obtained in the laboratory on 41 iron, nickel, copper, titanium, and molybdenum-base alloys in a Salton Sea KGRA-type brine at 232°C and pH 6.1. The general corrosion rates for these alloys in deaerated brines and brines containing carbon dioxide and methane were similar. The addition of oxygen to the brines sharply increased the incidence of stress corrosion cracking, crevice corrosion, and, except for titanium-base alloys, pitting. The most corrosion-resistant alloys in deaerated brine and brine containing carbon dioxide and methane were the high-chromium-molybdenum iron-base alloys (E-Brite 26-1), high-nickel alloys (Hastelloy S and Hastelloy C-276), titanium-base alloys (Ti-1.7W, Ti-2Ni, Ti-0.2Pd, and TiCode-12), and TZM. Only titanium-base alloys were resistant to corrosion in oxygenated brines. In deaerated brine, general, crevice, and pitting corrosion of carbon and low-alloy steels was relatively unaffected by pH in the range 6.1 to 3.0. Below pH 3.0, general and pitting corrosion rates increased sharply, and the alloys appeared more susceptible to crevice corrosion. Welds in carbon steel made with E6011, E6012, and E7018 rods corroded at rates similar to those of nonweld specimens and did not require heat treatment. Copper significantly lowered the resistance of low-alloy steels to general corrosion, whereas chromium, nickel, silicon, and titanium improved it.

In field tests, Type 1020 carbon steel, Type 4130 steel, and Types 410 and 430 stainless steels exhibited poor corrosion resistance and appear unsatisfactory for applications in high-temperature geothermal brine and steam environments. General corrosion and pitting rates for carbon and Type 4130 steel were high. Types 410 and 430 stainless steels pitted readily, and all four alloys were susceptible to crevice corrosion.

Acknowledgment

We gratefully acknowledge the generous contributions of alloys for inclusion in these studies by Airco Vacuum Metals Corp., Allegheny Ludlum Steel Corp., and Timet Corp., Pittsburgh, Pa.; Babcock and

Wilcox, Beaver Falls, Pa.; Cabot Corp., Kokomo, Ind.; Carpenter Technology Corp., Reading, Pa.; Huntington Alloy Products Division, International Nickel Corp., Huntington, W. Va.; RMI Co., Niles, Ohio; Sandvik, Inc., Scranton, Pa.; U.S. Steel Corp., Monroeville, Pa.; and the Albany Research Center, Bureau of Mines, Albany, Ore.

References

[1] Helgeson, H. C., *American Journal of Science,* Vol. 266, 1968, p. 129.
[2] Palmer, T. D., "Characteristics of Geothermal Wells Located in the Salton Sea Geothermal Field, Imperial Valley, California," UCRL-51976, University of California, Livermore, Calif., Dec. 1975.
[3] Carter, J. P. and McCawley, F. X., *Journal of Metals,* Vol. 30, March 1978, p. 11.
[4] Carter, J. P., McCawley, F. X., Cramer, S. D., and Needham, P. B., Jr., "Corrosion Studies in Brines of the Salton Sea Geothermal Field," Bureau of Mines Report of Investigations 8350, Bureau of Mines, Washington, D.C., 1979.
[5] Needham, P. B., Jr., Riley, W. D., Connor, G. R., and Murphy, A. P., *Society of Petroleum Engineers Journal,* Vol. 32, 1980, p. 105.
[6] Cramer, S. D. and Needham, P. B., Jr., "Linear Polarization Measurements at High Temperatures in Hypersaline Geothermal Brines," Bureau of Mines Report of Investigations 8308, Bureau of Mines, Washington, D.C., 1978.
[7] Barnes, H. L., Downs, W. F., Rimstidt, J. D., and Hayba, D. C., "Control of Silica Deposition in Geothermal Systems," Annual report on Bureau of Mines Grant G01551401, July 1977; available for consultation at Bureau of Mines Library in Avondale, Md. 20782.
[8] Riley, W. D., Cramer, S. D., Walters, R. P., and McCawley, F. X., *Society of Petroleum Engineers Journal,* to be published.
[9] Carter, J. P. and Cramer, S. D., in *Corrosion Problems in Energy Conversion and Generation,* C. S. Tedmon, Ed., Electrochemical Society, Princeton, N.J., 1974, pp. 240-250.
[10] Cramer, S. D. and Carter, J. P., Report of Investigations 8415, Bureau of Mines, Washington, D.C., 1980.
[11] Needham, P. B., Jr., Cramer, S. D., Carter, J. P., and McCawley, F. X., "Corrosion Studies in High-Temperature, Hypersaline Geothermal Brines," Paper No. 59 at Corrosion/79, National Association of Corrosion Engineers, Houston, Tex.
[12] Cramer, S. D., in *Corrosion Problems in Energy Conversion and Generation,* C. S. Tedmon, Ed., Electrochemical Society, Princeton, N.J., 1974, p. 251.
[13] Cramer, S. D., *Industrial and Engineering Chemistry—Process Design and Development,* Vol. 19, 1980, p. 675.
[14] Blackburn, M. J., Feeney, J. A., and Beck, T. R., "Stress-Corrosion Cracking of Titanium Alloys," in *Advances in Corrosion Science and Technology,* M. G. Fontana and R. W. Staehle, Eds., Vol. 3, Plenum Press, New York, 1973, p. 216.
[15] Copson, H. R., *Proceedings of the American Society for Testing and Materials,* Vol. 48, 1948, p. 591.
[16] Shannon, D. W., "Corrosion of Iron-Base Alloys Versus Alternate Materials in Geothermal Brines," PNL-2456, Pacific Northwest Laboratories, Richland, Wash., Nov. 1977.
[17] Syrett, B. C., MacDonald, D. D., Shih, H., and Wing, S. S., "Corrosion Chemistry of Geothermal Brines," Pt. 2, NSF(RANN) Grant No. AER 76-00713, National Science Foundation, Washington, D.C., Dec. 1977, pp. 16-18.
[18] Bishop, H. K., Bricarello, J. R., Enos, F. L., Hodgdon, N. C., Jacobson, W. O., Li, K. K., Mulliner, D. K., and Swanson, C. R., "SDG&E-ERDA Geothermal Loop Experimental Facility," SAN/1137-5, Department of Energy, Washington, D.C., March 1977.
[19] Bishop, H. K., Bricarello, J. R., Enos, F. L., Hanenburg, W. H., Hodgdon, N. C.,

Jacobson, W. O., Li, K. K., and Swanson, C. R., "SDG&E-DOE Geothermal Loop Experimental Facility," SAN/1137-8, Department of Energy, Washington, D.C., Oct. 1977.
[20] Sugimeto, K. and Sawada, Y., *Corrosion Science,* Vol. 17, 1977, p. 425.
[21] Hashimoto, K., Asami, K., and Teramoto, K., *Corrosion Science,* Vol. 19, 1979, p. 3.
[22] Posey, F. A., Palko, A. A., and Bacarella, A. L., "Corrosivity of Geothermal Brines," ORNL/TM-6159, Oak Ridge National Laboratory, Oak Ridge, Tenn., Nov. 1977.
[23] Goldberg, Alfred, "Comments on the Use of 316 L Stainless Steel Cladding at the Geothermal Niland Test Facility," UCID-17113, University of California, Livermore, Calif., 30 April 1976.
[24] Urbanek, M. W., Hornburg, C. D., and Lindal, B., "Research on Geothermal Mineral Extraction Complex: Phase 1—Preliminary Technical and Economic Assessment," U.S. BuMines Contract No. J0275091, DSS Engineers, Fort Lauderdale, Fla., Sept. 1978.

W. T. Lee[1] and D. Kramer[2]

Corrosion of Structural Steels in High-Salinity Geothermal Brine

REFERENCE: Lee, W. T., and Kramer, D., **"Corrosion of Structural Steels in High-Salinity Geothermal Brine,"** *Geothermal Scaling and Corrosion, ASTM STP 717,* L. A. Casper and T. R. Pinchback, Eds., American Society for Testing and Materials, 1980, pp. 142–154.

ABSTRACT: In support of geothermal power plant development, Rockwell International's Energy Systems Group (ESG) has conducted extensive studies and tests on the behavior of materials in geothermal environments. The materials tested include both metallic and nonmetallic materials. This paper presents Rockwell's recent stress corrosion test data on structural steels and their weldments under uniaxial loading (stress, 80 percent of yield strength) in static Sinclair 4 geothermal brine at 204°C for 1700 h (10 weeks). The alloys tested were 26Cr-1Mo, 18Cr-2Mo, 9Cr-1Mo, and Type 316 stainless steel and their weldments. Under these test conditions, the following results were obtained:

1. In testing without air contamination: (*a*) Alloys 26Cr-1Mo, 18Cr-2Mo, and 9Cr-1Mo and their weldments were resistant to stress corrosion cracking; and (*b*) the average corrosion rates for the welded specimens made of 26Cr-1Mo, 18Cr-2Mo, and 9Cr-1Mo were 7.7 µm/year, 7.4 µm/year, and 18.2 µm/year, respectively.
2. In testing with air contamination: (*a*) the 18Cr-2Mo and 9Cr-1Mo ferritic steels were found to be resistant to stress corrosion cracking in air-contaminated brine; (*b*) Alloy 26Cr-1Mo and Type 316 stainless steel exhibited cracking typical of stress corrosion; and (*c*) the average corrosion rates were much higher than those obtained for the alloys without air contamination.

Results of the posttest mechanical properties testing and metallographic analysis are also presented.

KEY WORDS: fracture, stress corrosion, steels, geothermal brine, scaling

Electric power generation using geothermal brines offers great promise as an alternative power source, but its realization requires that a number of severe technical problems be solved, such as selection of an economic

---

[1] Member of the technical staff, Research and Technology Department, Rockwell International, Energy Systems Group, Canoga Park, Calif.
[2] Manager, Research and Technology Department, Rockwell International, Energy Systems Group, Canoga Park, Calif.

power conversion cycle and control of scaling and corrosion in plant components [1].[3]

Because of the high temperatures, low pH, high salinity, and high concentration of aggressive impurities in the brines, most of the common construction materials suffer severe corrosion and erosion problems in most brines and steam from geothermal wells in the Imperial Valley, Calif. [1,2,3,4]. Corrosion encompasses general uniform corrosion, pitting and crevice corrosion, stress corrosion or sulfur stress cracking, and hydrogen embrittlement. In addition, the presence of entrained solid particles in the brine and steam increases material removal due to erosion. Accordingly, selection of materials resistant to this aggressive environment is paramount to reliable operation of geothermal systems. Rockwell International, Canoga Park, Calif., has conducted extensive studies and tests on the behavior of materials in liquid-dominated geothermal environments, using the geothermal fluids from the Sinclair 4 well in the Imperial Valley, Calif. The materials tested included both metallic and nonmetallic materials. The earlier test results have been presented in Ref 2.

The purview of this paper is limited to the more recent corrosion test data on structural alloys and their weldments in static Sinclair 4 brine at 204°C. The Sinclair 4 well is located in the Salton Sea known geothermal field, in Calif., a liquid-dominated high-temperature (bottom hole temperature, 260°C) geothermal reservoir with high salinity and high total dissolved solids (~30 percent), displayed in Table 1. The noncondensable gases from Sinclair 4 brine were analyzed to contain (on a dry basis) 91 percent carbon dioxide (CO_2), 7 percent nitrogen (N_2) 1.5 percent hydrocarbon (HC), and trace amounts of hydrogen sulfide (H_2S) and ammonia (NH_3). The Sinclair 4 well and test system are shown in Fig. 1.

Experimental Procedure

Testing was conducted in Teflon-lined retorts filled with geothermal brine. Each retort contained about 400 ml of geothermal brine, together with three prestressed tensile specimens (in three loaders) submerged in the brine. Three thermocouples were installed to continuously record the temperature of each specimen during the test. A temperature deviation of ±2°C was obtained during each test. As shown in Fig. 2, the prestressed tensile specimens had a nominal size of 0.32 by 7.0 by 0.24 cm and were fabricated to the requirements of ASTM Recommended Practice for Preparing, Cleaning, and Evaluating Corrosion Test Specimens (G 1-72). The steels tested were Alloys 26Cr-1Mo (E-Brite 26-1), 18Cr-2Mo, and 9Cr-1Mo and Type 316 stainless steel and their weldments. The composition of the tested alloys is shown in Table 2. The wrought specimens were tested in an

---
[3] The italic numbers in brackets refer to the list of references appended to this paper.

TABLE 1—*Concentrations of components of Sinclair 4 brine.*[a]

Component	Concentration, ppm
Sodium	62 600 to 80 500
Potassium	14 000 to 19 100
Lithium	192 to 257
Barium	1 200 to 1 630
Calcium	25 700 to 35 200
Strontium	513 to 640
Magnesium	50 to 113
Boron	100 to 270
Silica	415 to 675
Iron	1 130 to 1 680
Manganese	1 470 to 1 820
Lead	80 to 117
Zinc	322 to 425
Copper	2.4 to 4.4
Silver	0.35 to 0.55
Chloride	130 725 to 172 200
ΣCO_2	662 to 1 248
ΣS	182 to 207
Total dissolved solids	238 500 to 314 600

[a] After flashing.

"as-machined" condition, whereas the weldments were stress relieved prior to testing except for the welded 18Cr-2Mo steel specimens, which were tested in an "as-welded" condition. The welding had been performed by the tungsten inert gas (TIG) process. Specimen prestressing to 80 percent of the yield strength at room temperature was accomplished in an Instron tensile machine. To eliminate any galvanic coupling between dissimilar alloys, the loaders were fabricated of the same alloy as the specimen material.

Brine loading into the retorts was accomplished in a glove box at ambient temperature under a slightly positive argon atmosphere (99.995 percent pure argon). The absence of oxygen in the glove box was verified by a zero reading on a calibrated oxygen meter. The pH of the brine solution was measured and recorded (pH = 4). The brine solution was loaded into the retort to approximately 3 mm (⅛ in.) from the top of the retort. After closure of the retort and verification of sealing, the retorts were lowered into a three-zone furnace for testing at 204°C. All testing was performed for a period of 10 weeks. During the test, each retort was periodically monitored for brine leakage. No leakage was found except in three early retorts with wrought specimens fabricated of Alloys E-Brite 26-1, 18Cr-2Mo, and 9Cr-1Mo and Type 316 stainless steel. After achievement of their 10-week exposure time, these three retorts were opened, and the seals were found to be damaged. The brine had evaporated, leaving salt deposits on

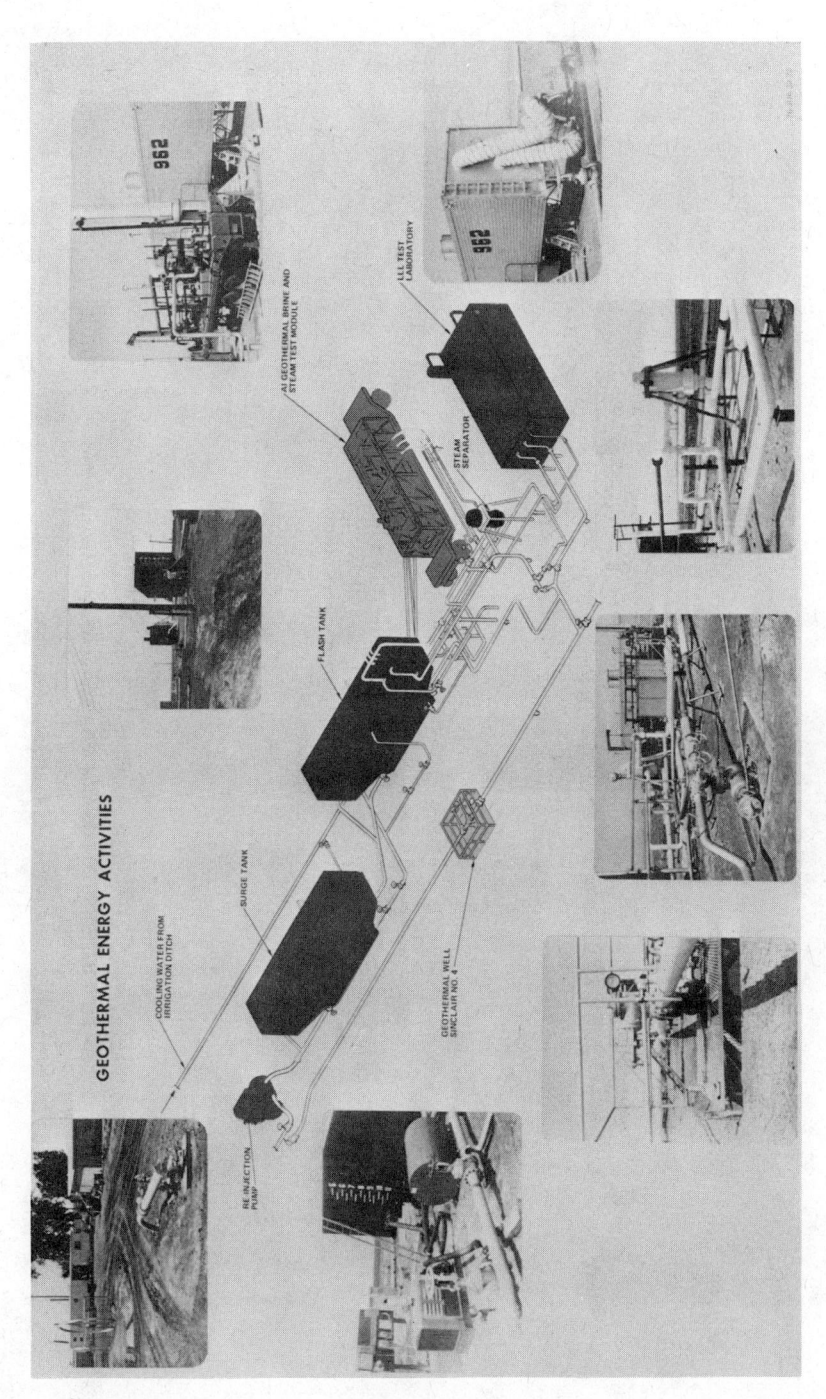

FIG. 1—*Sinclair 4 geothermal well and test system.*

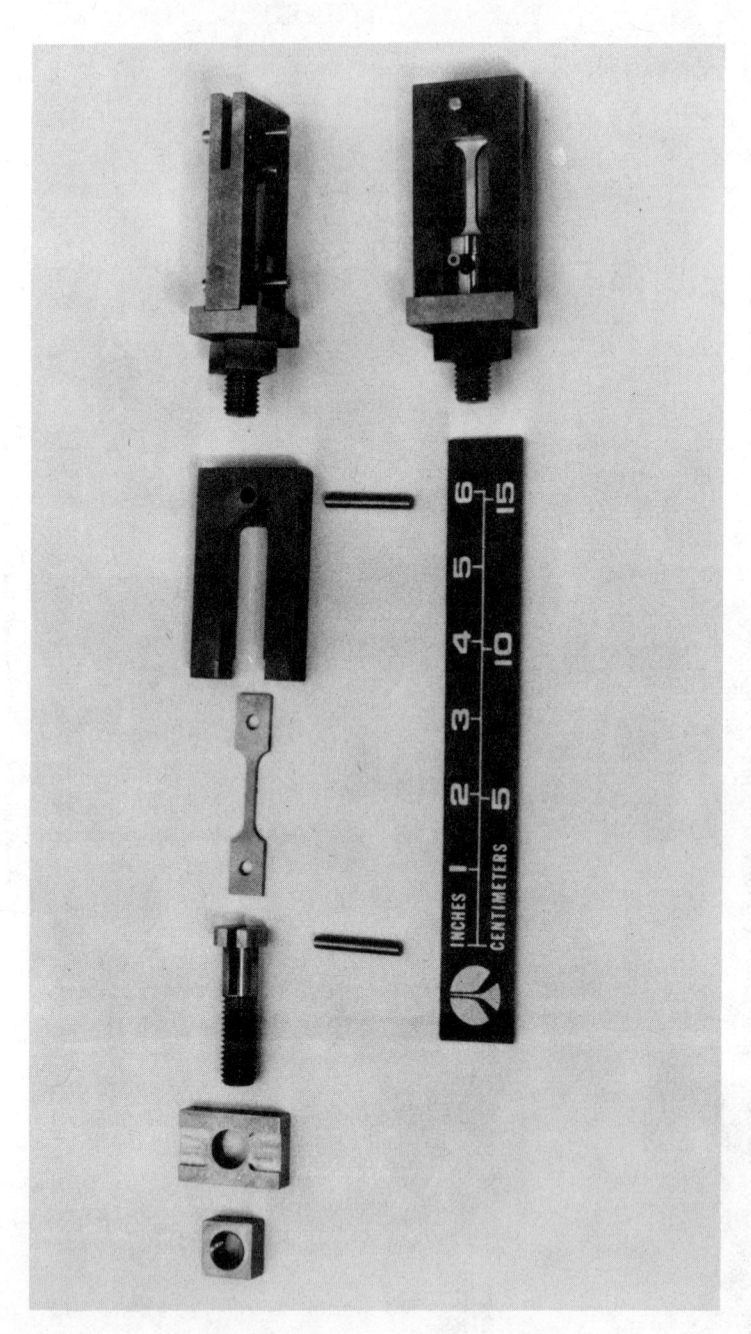

FIG. 2—*Stress corrosion test apparatus in geothermal brine.*

TABLE 2—*Compositions of alloys, by weight percent.*

Composition	Type 316 Stainless Steel (Annealed Typical)	18Cr-2Mo, (Annealed, Typical)	E-Brite 26-1 (Heat No. 1390B)	9Cr-1Mo (Heat No. 536427-3)
C	0.08 max	C + N = 0.04 max	<10 ppm	0.084
Mn	2.0 max	1.0 max	0.02	0.52
P	0.045 max	0.04 max	0.010	0.019
S	0.03 max	0.03 max	0.009	0.016
Si	1.0 max	1.0 max	0.23	0.44
Cr	17.4	18.1	26.1/25.9	9.05
Ni	13.2	0.4 max	0.11	...
Cb		...	...	...
Ta		...	...	...
Mo	2.17	2.4	0.99/1.1	1.00
Ti	...	0.32	...	...
Cu		0.2 max	0.01	...
Co		...	...	...
N		C + N = 0.04 max	0.010	...

the specimen and loaders. All other tests were completed successfully without any leakage. After exposure, the specimens were cleaned, weighed, and examined for corrosion and cracks. The uncracked specimens were again tested in tension to determine the effect of brine exposure on the short-term mechanical properties.

Results and Discussion

The results of the 10-week tests at 204°C (400°F) are summarized in Table 3, which shows that all the as-machined E-Brite 26-1 and Type 316 stainless steel (for control) cracked in the retort that leaked, presumably due to air contamination, since oxygen is required for stress corrosion of Type 316 stainless steel [5]. No cracking was found in the as-machined wrought 9Cr-1Mo and 18Cr-2Mo steels in the same environment. In addition, no cracking was found in the as-machined and welded 18Cr-2Mo, stress-relieved and welded E-Brite 26-1, and welded 9Cr-1Mo steels in brine without air contamination.

The microstructures of the cracked Type 316 stainless steel and E-Brite 26-1 alloy in air-contaminated brine are shown in Figs. 3 and 4, respectively. The microstructures displayed transgranular and intergranular cracking modes with branching characteristics typical of stress corrosion. It is normal to find stress corrosion cracking of Type 316 stainless steel in a high-temperature acidic chloride environment; it was, however, a surprise to find the branching cracks in all the E-Brite 26-1 specimens tested in air-contaminated brine, since E-Brite 26-1 is reported to be resistant to stress corrosion in severe chloride environments [6]. This is the first reported

TABLE 3—*Stress corrosion testing results in Sinclair 4 brine for 10 weeks at 204°C (400° ± 5°F).*[a]

Retort No.	Specimen Material	Stress,[b] MPa, (ksi)	Visual Examination
1[c]	18Cr-2Mo	324 (47)	N.F.[d]
	18Cr-2Mo	283 (41)	N.F.
	316 stainless steel	241 (35)	cracked
2[c]	26Cr-1Mo	324 (47)	cracked
	26Cr-1Mo	324 (47)	cracked
	316 stainless steel	276 (40)	cracked
3[c]	9Cr-1Mo	310 (45)	N.F.
	9Cr-1Mo	276 (40)	N.F.
	316 stainless steel	241 (35)	cracked
4	Welded 26Cr-1Mo	248 (36)	N.F.
	Welded 26Cr-1Mo	248 (36)	N.F.
	Welded 26Cr-1Mo	248 (36)	N.F.
5	Welded 18Cr-2Mo	269 (39)	N.F.
	Welded 18Cr-2Mo	269 (39)	N.F.
	Welded 18Cr-2Mo	269 (39)	N.F.
6	Welded 9Cr-1Mo	331 (48)	N.F.
	Welded 9Cr-1Mo	331 (48)	N.F.
	Welded 9Cr-1Mo	331 (48)	N.F.

[a] pH of Sinclair 4 brine = 4.
[b] Specimens were stressed to 80 percent of yield strength at room temperature.
[c] Air contaminated.
[d] N.F. = no failure or cracking.

incidence of cracking of wrought E-Brite 26-1 in a geothermal brine environment. In previous work with this material in Salton Sea brine, only pitting and corrosion were experienced [3,4,7,8]; no stress corrosion cracking was ever reported on this alloy in a geothermal environment. However, an electron-beam-melted 26Cr-1Mo (E-Brite 26-1) steel, in wrought or welded form, was found to be highly susceptible to stress corrosion cracking in a thermally cycled (430°C to 280°C) steam environment with 10 ppm chlorine and 20 ppm oxygen [9].

Microstructural examination of the failed specimens did not reveal any grain boundary precipitate that could be associated with stress corrosion. The presence of oxygen in the brine not only caused cracking of the E-Brite 26-1 steel but also accelerated the corrosion rates of the steels, as shown in Table 4; this result is consistent with the data reported in the literature [7,8].

Typical microstructures of the 9Cr-1Mo steel specimens are shown in Fig. 5, which shows pitting of the steel. Based on the maximum pit depth, the maximum corrosion or penetration rate is 385 μm/year (15 mils per year), which is more than 20 times higher than the corrosion rate of 18.2 μm/year (0.7 mils per year) calculated by weight loss, as shown in Table 4. The importance of pitting corrosion cannot be overemphasized in com-

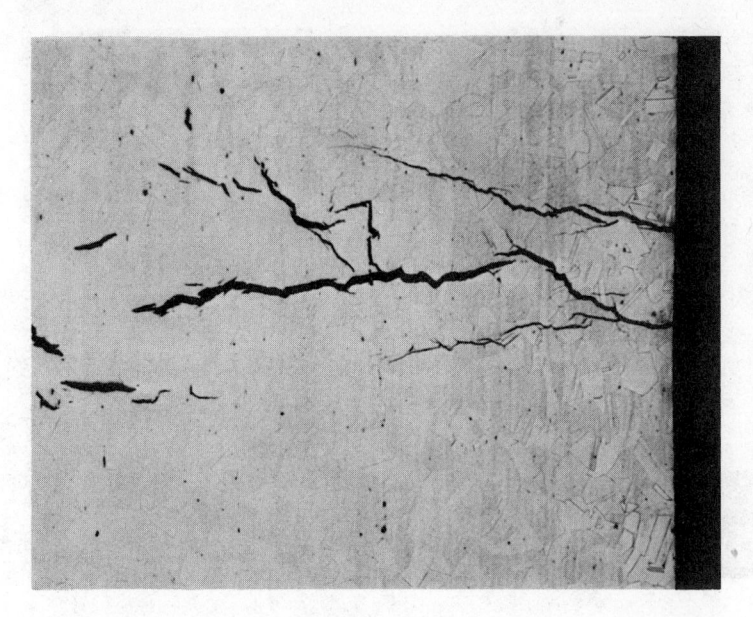

FIG. 3—Microstructures (×100) of Type 316 stainless steel: (left) tested in brine, (right) as-machined specimen.

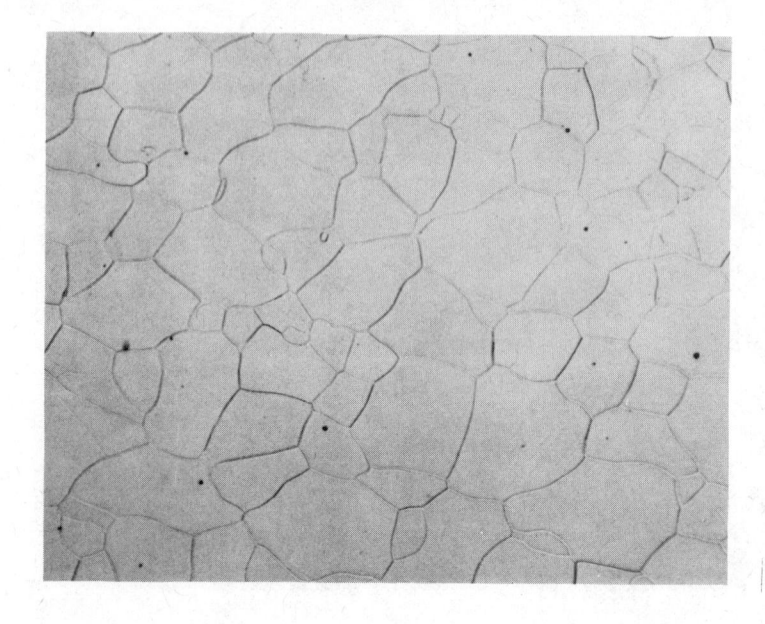

FIG. 4—Microstructure (×100) of 26Cr-1Mo steel specimens: (left) 26Cr-1Mo steel specimen after exposure to brine, (right) as-machined 26Cr-1Mo.

TABLE 4—*Effect of air contamination in brine on corrosion rates of alloys.*

	Corrosion Rate, μm/year (mils per year)[a]	
Alloy	No Air Contamination	Air Contamination[b]
9Cr-1Mo	18.2 (0.7)	1144 (44)
18Cr-2Mo	7.4 (0.29)	910 (35)
26Cr-1Mo	7.7 (0.30)	1040 (40)
Type 316 stainless steel	...	988 (38)

[a] Based on weight change average of three specimens.
[b] Calculation is based on total exposure time.

ponents for geothermal applications, since localized corrosion failures of components are prevalent [1,2,3]. The design of components for geothermal applications should incorporate corrosion allowances based on maximum pit depth instead of on weight loss.

The uncracked specimens were subjected to tension testing at room temperature to determine the effect of geothermal brine exposure on the short-term mechanical properties. As shown in Table 5, the exposed specimens displayed a slightly higher yield strength, comparable tensile strength, and a slightly reduced elongation. Under the test conditions, the brine environment apparently had no significant effects on the short-term mechanical properties. The increase in yield strength and reduction in elongation are due to material work-hardening as a result of deformation under tension loading and possibly of thermal aging at 204°C. Metallographic examination of the fractured surfaces revealed cup-and-cone ductile deformation without any apparent adverse effect of corrosion or embrittlement. However, it is expected that longer exposure would produce increased pitting in 9Cr-1Mo steel, which could produce a stress-concentrating effect at the pits, thus reducing the load-bearing capability. It should be noted that, for stress corrosion cracking test results without knowledge of the exact failure mechanism, a failure in an alloy corroborates the alloy's positive vulnerability to stress corrosion cracking; but the converse is not necessarily true, since many service failures occur only after a long exposure period.

Conclusions

Based on the results of testing for 10 weeks in Sinclair 4 geothermal brine at 204°C (400°F), the authors concluded that:

1. Alloys 9Cr-1Mo, 18Cr-2Mo, and 26Cr-1Mo (E-Brite 26-1) and their

FIG. 5—*Typical microstructures of 9Cr-1Mo steel before and after brine exposure: (left) unexposed 9Cr-1Mo steel (×250), (right) brine-exposed 9Cr-1Mo steel (×250) showing pits [maximum pit depth, 51 μm (2 mils)].*

TABLE 5—*Effects of brine exposure on room temperature mechanical properties of 9Cr-1Mo, 18Cr-2Mo, and 26Cr-1Mo steels and their weldments.*

Alloy	Yield Strength MPa	ksi	Tensile Strength MPa	ksi	Elongation, percent
\multicolumn{6}{c}{Room Temperature, 70°F, Air}					
18Cr-2Mo	407	59	552	80	39
9Cr-1Mo	386	56	621	90	37
E-Brite 26-1	409	59	532	77	40
Welded 9Cr-1Mo[a]	414	60	621	90	10
Welded 18Cr-2Mo[b]	331	48	517	75	18
Welded 26Cr-1Mo[c]	310	45	462	67	12.4
\multicolumn{6}{c}{Room Temperature, After Exposure to Brine[d]}					
18Cr-2Mo	428	62	559	81	24
9Cr-1Mo	455	66	614	89	24
E-Brite 26-1	[e]	[e]	[e]	[e]	[e]
Welded 9Cr-1Mo[a]	434	63	614	89	9
Welded 18Cr-2Mo[b]	386	56	524	76	20
Welded 26Cr-1Mo[c]	345	50	469	68	13

[a] Stress-relieved at 732°C (1350°F) for 20 min, air cooled.
[b] As-welded condition.
[c] Stress-relieved at 871°C (1600°F) for 10 min., water quenched.
[d] Prestressed tensile specimen exposed to Sinclair 4 brine for 10 weeks at 400°F.
[e] Broken due to stress corrosion cracking during test.

stress-relieved weldments were resistant to corrosion and stress corrosion cracking in brine.

2. Alloy E-Brite 26-1 and Type 316 stainless steel suffered cracking typical of stress corrosion in air-contaminated brine.

3. Air contamination accelerated corrosion of the steels in brine.

4. For design of components for geothermal applications, the corrosion allowance should be based on maximum penetration due to localized corrosion (such as pitting) and not estimated by weight loss.

5. Brine exposure did not affect the short-term mechanical properties of the resistant alloys.

Acknowledgment

This work was supported by Rockwell International Corp. Independent Research and Development funds. Appreciation is extended to Climax Molybdenum Co., Division of AMAX, Torrence, Calif., and Airco Vacuum Metals Co., Berkeley, Calif., for supplying some of the 18Cr-2Mo and E-Brite 26-1 specimens, respectively.

References

[1] Hajela, G. P., Lee, W. T., Recht, H. L., Schapiro, A. R., and Springer, T. H., "Analysis of Selected Geothermal Power Cycles," presented at the National Geothermal Conference, Palm Springs, Calif., 19-22 April 1976.
[2] Recht, H. L., Lee, W. T., Springer, T. H., "Evaluation of Corrosion in a Geothermal Well Liner," Paper 118, presented at the Electrochemical Society, 150th Fall Meeting, Las Vegas, Nev., 17-22 Oct. 1976.
[3] Carter, J. P. and Cramer, S. D., *Corrosion Problems in Energy Conversion and Generation,* 1974, p. 240.
[4] Carter, J. P., McCawley, F. X., Cramer, S. D., and Needham, P. B., Jr., "Corrosion Studies in Brines of the Salton Sea Geothermal Field," USBM-RI8350, U.S. Bureau of Mines, College Park, Md., 1979.
[5] Williams, W. L., *Corrosion,* Vol. 13, No. 8, 1957.
[6] Knoth, R. J., *Electric Furnace Proceedings,* Vol. 29, 1971, p. 123.
[7] McCawley, F. X., Carter, J. P., and Cramer, S. D., "Materials for Use in Corrosive Geothermal Brine Environments," presented at the U.S. Department of Energy, Division of Geothermal Energy, Symposium on Materials in Geothermal Energy Systems, Austin, Tex., 23-25 May 1978.
[8] McCright, R. D. and Garrison, R. E., "Corrosion Testing in Mulitple-Stage Flash System," presented at the U.S. Department of Energy, Division of Geothermal Energy, Symposium on Materials in Geothermal Energy Systems, Austin, Tex., 23-25 May 1978.
[9] Hammond, J. P., Patriarca, P., Slaughter, G. M., and Maxwell, W. A., *Materials Performance,* Vol. 14, No. 11, Nov. 1975, pp. 41-52.

C. Arnold, Jr.,[1] K. W. Bieg,[2] and J. A. Coquat[3]

Degradation of Elastomers in Geothermal Environments

REFERENCE: Arnold, C., Jr., Bieg, K. W., and Coquat, J. A., **"Degradation of Elastomers in Geothermal Environments,"** *Geothermal Scaling and Corrosion, ASTM STP 717,* L. A. Casper and T. R. Pinchback, Eds., American Society for Testing and Materials, 1980, pp. 155-163.

ABSTRACT: The thermochemical degradation and sealing capability of various commercial O-rings in geothermal brine was evaluated at elevated temperatures and pressures to facilitate the selection of suitable elastomeric seals for the cable heads of geothermal logging tools. The extent of degradation was determined by measuring the change in tensile properties and glass transition temperatures that occurred as a result of exposure to brine at 548 K (275°C) and 34 MPa (5000 psi). The sealing capabilities of the O-rings in brine under these conditions were evaluated, using O-ring fixtures that simulated the cable head of a commercial logging tool. All of the elastomers became degraded to some extent. The elastomers that retained some measure of elasticity and also were not breached by brine in seal tests were the perfluoroelastomers, ethylene-propylene rubbers, and a peroxide-cured fluoroelastomer. Data obtained on the properties of aged specimens of these elastomers suggest that the most likely mode of degradation was chain scission. Nitrile rubber, silicone, and the peroxide-cured fluoroelastomer suffered a complete loss of elasticity and failed as seals. Data obtained on the properties of these elastomers indicate that extensive cross-linking took place. Polychloro-*p*-xylylene coatings were found to inhibit the rate of degradation in both brine and steam/sour gas environments.

KEY WORDS: thermochemical degradation, elastomers, geothermal brine, chain scission, cross-linking, coatings, seals, perfluoroelastomers, ethylene-propylene rubbers, silicones, fluoroelastomers, nitrile rubbers, scaling, corrosion

Elastomers that are resistant to the combined effects of steam, brine, and hydrogen sulfide (H_2S) are needed as seals in geothermal applications. Typical seal applications include packers, O-rings for logging tool cable heads, and ram seals for blowout preventers. Although elastomeric seals are generally preferred over metal seals in these applications, because they are

[1] Chemist, Sandia Laboratories, Albuquerque, N. Mex. 87115.
[2] Chemist, Sandia Laboratories, Albuquerque, N. Mex. 87115.
[3] Chemist, Sandia Laboratories, Albuquerque, N. Mex. 87115.

easy to install and relatively tolerant of irregular surfaces, they frequently fail as a result of environmental stresses that exist in a geothermal borehole. The chief environmental stresses in a geothermal borehole are: (1) high temperatures (432 to 633 K [150 to 360°C]), (2) high pressures [up to 48 MPa (7000 psi)], and (3) corrosive well fluids (water, steam, brine, carbon dioxide, and hydrogen sulfide). From a chemical standpoint, these conditions are hydrolytic and reducing in nature. Under these conditions, elastomers can fail by: (1) extrusion due to heat and mechanical stress [1,2],[4] (2) mechanical weakness at high temperatures and pressures [3], and (3) thermochemical degradation [1-4]. By and large, all three of these factors work simultaneously and perhaps synergistically to cause seal failure. In the present study, we evaluated the thermochemical stability of O-ring seals for potential use in a geothermal logging tool and investigated the feasibility of enhancing the environmental resistance of elastomers by means of coatings. The use of coatings for these applications had not been previously investigated. Furthermore, our studies were carried out with a nonsynthetic brine and included several classes of rubbers not evaluated by others.

Procedures

The environmental stability of the eleven commercially available O-rings listed in Table 1 was determined by exposure to a geothermal brine obtained from the Salton Sea basin (Mesa Well No. 6-1) at 548 K (275°C) and 34.5 MPa (~5000 psi) for 8.647×10^4s (24 h). The salinity of this brine, which contained no hydrogen sulfide, was 0.0158 kg NaCl/kg (15 800 ppm). These runs were carried out in a bolted-closure pressure vessel equipped with a tubular furnace and a temperature controller.

The extent of degradation was determined from changes that occurred in such mechanical and thermomechanical properties as tensile strength, compression set, and glass transition temperature. The tensile properties and compression set properties of both virgin and aged O-rings were determined by the ASTM Standard Methods of Testing Rubber O-Rings (D 1414-78) and ASTM Tests for Rubber Property—Compression Set [D 395-69 (1975)]. The following deviations from the standardized protocol were made: (1) in the tensile test the O-ring was cut; (2) aging in brine necessitated the use of longer cooling periods than those called for in ASTM Method D 395-69 (1975); and (3) the spacer size was adjusted to give a deflection of 22 percent rather than 25 percent, as recommended in ASTM Method D 395-69 (1975). The glass transition temperature was determined using a DuPont thermomechanical analyzer (Model 943) run in the penetration mode with a 0.050-kg weight at a heating rate of 0.33 K/s (20°C/min).

The sealing efficiency of O-rings under simulated borehole conditions was

---

[4]The italic numbers in brackets refer to the list of references appended to this paper.

TABLE 1—*Commercial O-rings evaluated in brine tests.*

O-Ring Designation	Generic Type	Supplier	Ingredient Variations
Kalrez[a] (1050, 3074)	perfluoroelastomer	DuPont	3074 contains asbestos
E692-75	ethylene-propylene-diene-monomer (EPDM)	Parker Seal	
E540-80	ethylene-propylene copolymer (EPR)	Parker Seal	
V747-75	fluoroelastomer	Parker Seal	E60C (DuPont) or Fluorel 2170 (3M) gumstocks, cured by bisphenol
V709-90	fluoroelastomer	Parker Seal	E60C (DuPont) or Fluorel 2170 (3M) gumstocks, cured by bisphenol
V752-7	fluoroelastomer	Irving Moore	Viton GH cured
VW-211V	fluoroelastomer	A. W. Chesterton	elastomer coated with 3.81×10^{-4} m (15 mil) of moldable Teflon
N552-90	nitrile rubber	Parker Seal	
S604-70	silicone	Parker Seal	
L449-70	fluorosilicone	Parker Seal	fluorosilicone gumstock

[a]Trade name of DuPont.

determined using the two-part fixture shown in Fig. 1. This fixture, which simulates the cable head of a geothermal logging sonde, has a diametral gap of 0.00076 m (0.030 in.), which results in a deflection of the O-ring of 22 percent. Molecular sieves (Linde 5A, made by Union Carbide) were placed in the female part of the two-part fixture to absorb any water vapor that penetrated through the O-ring. After assembly, the entire fixture was heated in brine at 548 K (275°C) for 8.647×10^4s (24 h). After completion of the run, the moisture content of the molecular sieves was determined using a moisture analyzer (Model 26-321 by Consolidated Electrodynamics Corp.).

To enhance their environmental resistance, the following O-rings were coated with a film of Parylene C[5] (polychloro-*p*-xylylene) 3.2×10^{-5}m (0.00125 in.) thick via pyrolytic vapor phase deposition: Specimens of V747-75, V709-90, E692-75, E540-80, N552-90, S604-90. The aging of these coated O-rings was accomplished with the same procedure used to age the uncoated O-rings. The extent of degradation was assessed by determining the glass transition temperature of the underlying rubber before and after aging.

Results

Data on the deterioration of mechanical and thermomechanical properties of the uncoated elastomers as a result of exposure to brine at 548 K (275°C) and 34 MPa (~5000 psi) for 8.647×10^4s (24 h) are compiled in Table 2. The elastomers in Table 2 are listed in order of decreasing stability during exposure to brine. It is evident from these data that the most brine-resistant elastomers were the perfluoroelastomers, the ethylene-propylene copolymer, the ethylene-propylene-diene monomer (EPDM) rubber, and the peroxide-cured fluoroelastomer; the least stable elastomers included the silicones, the nitrile rubber, and the bisphenol-cured fluoroelastomer. These data also indicate that there is a good correlation between the thermochemical stability of the elastomer and its sealing ability. Thus, the five O-rings that passed the sealing efficiency test either showed no change or only slight changes in their glass transition temperatures. In contrast, the five O-rings that failed the seal efficiency test showed significant increases in their glass transition temperatures. Furthermore, the tensile properties of those elastomers that failed the seal efficiency test were untestable because of embrittlement or complete loss of mechanical integrity. Although the elastomers that passed the seal efficiency test suffered a significant loss of their tensile properties, they retained some elasticity when handled manually.

It should be noted that the tensile measurements were made on elastomers that had been completely immersed in the brine and thus had received a higher exposure to the brine than O-rings that were in the seal efficiency fixture. It also should be noted that failure in the seal efficiency test is defined

---

[5] Union Carbide trade name.

FIG. 1—*Photograph of a seal efficiency fixture.*

TABLE 2—*Degradation of uncoated elastomers in the presence of geothermal brine at 548 K/34 MPa/8.64 × 10⁴s (275°C/5000 psi/24 h).*

Material Designation	Generic Type	Change in Tensile Strength, %	Change in Elongation, %	Change in T_g^a, deg C	Seal Efficiency Test
Kalrez	perfluoroelastomer				
1050		−24	−6	0	pass
3074		no test	no test	0	pass
E692-75	ethylene-propylene-diene-monomer (EPDM)	−70	+23	0	pass
E540-80	ethylene-propylene copolymer (EPR)	−62	+30	+2	pass
V752-7	fluoroelastomer (peroxide cured)	−52	−68	+2	pass
V709-90	fluoroelastomer (bisphenol cured)	too brittle to obtain tensile properties		+12	fail
V747-75	fluoroelastomer (bisphenol cured)	too brittle to obtain tensile properties		>+140	fail
N552-90	nitrile rubber	too brittle to obtain tensile properties		+69	fail
S609-70	silicone	too brittle to obtain tensile properties		>+100	fail
L449-70	fluorosilicone	too brittle to obtain tensile properties		>+100	fail

[a] T_g = glass transition temperature.

as a breach in the seal, through which substantial amounts of liquid brine entered the fixture. Small quantities of moisture (~9 mg), however, were detected in the molecular sieves in those runs in which the O-rings passed the seal efficiency test. This is not surprising considering there was a ~34 MPa (~5000 psi) differential pressure across these seals, and the temperature was 548 K (275°C). It is not known whether or not this quantity of moisture can be tolerated in a geothermal logging tool.

The compression set results were worse than anticipated; all of the O-rings evaluated had compression sets of 100 percent or higher, which means that, in addition to taking a set, they also shrank. Correlation of the compression set results with functional performance in the seal efficiency tests was not good. On the basis of the compression set results, one would not have predicted that any of these elastomers would have passed the seal efficiency test. Here again, however, a relatively higher fraction of the rubber surface was exposed to the brine in the compression set test than in the seal efficiency test.

The results obtained in enhancement of the thermochemical stability of coatings of polychloro-p-xylylene and copolytetrafluoroethylene-perfluoropropylene (FEP) are summarized in Table 3. Except for the nitrile rubber (N552-90), the extent of degradation, as measured by the increase in the glass transition temperature of the coated elastomers, was less than that observed for the uncoated elastomers. Since coatings cannot prevent the breakdown of elastomers due to heat, it was not surprising that no improvement was shown by the nitrile rubber. By thermogravimetric analysis, it was shown that the nitrile rubber was the least thermally stable of all the elastomers (decomposition temperature, ~150°C).

Discussion

Except for the perfluoroelastomers, which extruded slightly but did not fail, extrusion did not play a prominent role in the failure of the rubber seals in a brine environment. The primary cause of failure was thermochemical degradation. If the loose network structure of an elastomer is broken down by random chain scission, the elastomer will lose its elasticity and will also suffer a loss of its tensile properties [5]. When extensive cross-linking occurs, the glass transition temperature of the rubber rises, and the rubber becomes both weak and brittle [5,6]. Random chain scission appeared to be the predominant process that occurred with the perfluoroelastomers, ethylene-propylene rubber, EPDM rubber, and peroxide-cured fluoroelastomer. These rubbers suffered a loss of tensile properties but retained some measure of elasticity. Furthermore, the glass transition temperatures of these rubbers was essentially unchanged after exposure to brine, and O-ring seals made from these formulations did not fail. Apparently, the extent of chain scission was not sufficient to affect the glass transition temperature of the rubber. On

TABLE 3—*Effect of polychloro-p-xylylene[a] and molded copoly tetrafluoroethylene-perfluoropropylene[b] coatings on the thermochemical stability of elastomers 548 K/34 MPa/8.64 × 10⁴s (275°C/5000 psi/24 h).*

Material Designation	Coating	Generic Class	Glass Transition Temperatures (Tg), deg C		
			Control	Uncoated, Change in deg C	Coated, Change in deg C
V709-90	polychloro-p-xylylene	fluoroelastomer	−23	−11(+12)	−20(+3)
V747-75	polychloro-p-xylylene	fluoroelastomer	−18	>+140(>+158)	−18(0)
VW211V	copoly (tetrafluoroethylene-perfluoropropylene) and propylene	fluoroelastomer	−21	>+140(>+158)	−16(+5)
N552-90	polychloro-p-xylylene	nitrile (Buna)	−29	+69(+98)	+78(+107)
S604-70	polychloro-p-xylylene	silicone	−127[c]	(>+100)[c]	−90(+37)[c]

[a] Union Carbide trade name is Parylene C; film thickness = 0.037 mm (0.00125 in.).
[b] DuPont trade name is Teflon FEP; film thickness = 0.38 mm (0.015 in.).
[c] Determined by differential scanning calorimetry (Perkin Elmer, DSC II).

the other hand, cross-linking seemed to predominate during brine exposure in the case of the bisphenol-cured fluoroelastomer, nitrile rubber, and silicones. The glass transition temperatures of these rubbers increased substantially. In fact, the rubbers became so embrittled that their tensile properties could not be obtained.

Elastomers that have labile, pendant groups have a tendency to undergo chain stripping [7]. Fluoride ion analyses revealed that fluoroelastomers became degraded in this manner after exposure to brine at elevated temperatures and pressures. In this case, the pendant labile groups were hydrogen and fluorine. Chain stripping seldom occurs exclusively; this process is usually accompanied by cross-linking and chain scission [8].

The stability order observed in these studies can be rationalized on the basis of well-known structure stability relationships. Generally, carbon-fluorine bonds are stronger than the carbon-hydrogen bonds [9]; therefore, it is not surprising that perfluoroelastomers, in which fluorine has been substituted for all the hydrogen in the backbone of the chain, are thermally and chemically more stable than hydrocarbon-type elastomers (that is, ethylene-propylene and EPDM rubbers) that contain no carbon-fluorine bonds or fluoroelastomers that have most but not all of their hydrogens replaced by fluorines. The latter class of elastomer degrades by chain stripping, as mentioned above. The superiority of EPDM rubber's brine resistance to that of fluoroelastomers may be attributable to the inherent incompatibility on the nonpolar backbone of the EPDM rubber with a high polar species such as water. The improved steam resistance of the peroxide-cured fluoroelastomer, in comparison with fluoroelastomers cured with phenolic-type curing agents, has been attributed to the higher hydrolytic stability of a carbon-carbon bonded cross-link in the peroxide-cured formulations than that of a carbon-oxygen bonded cross-link of the phenol-cured product [10]. The silicones became extensively degraded in brine because of their relatively poor hydrolytic stability [11]. Lack of thermal stability probably accounts for the failure of nitrile rubber in brine.

The enhanced brine stability of coated over uncoated elastomers is probably attributable to the ability of the coating to act as a barrier to limit the rate of diffusion of chemically reactive components of the environment into the bulk of the elastomers. The utility of coated elastomers as seals has not yet been demonstrated.

Conclusions

The chief mode of seal failure for elastomeric O-rings appears to be thermochemical degradation rather than extrusion. O-rings formulated from fully fluorinated gumstocks and hydrocarbon-based gumstocks (such as EPDM rubbers) exhibited the highest resistance to brine under simulated borehole conditions and should receive first consideration as seals in geothermal applications. Polychloro-*p*-xylylene coatings enhanced the resistance of some elastomers toward brine. Further experiments are needed to demonstrate the utility of coated O-rings in geothermal seal applications.

Acknowledgments

This work was supported by the U.S. Department of Energy under Contract DE-AC04-76-DP00789.

References

[1] Hirasuna, A. R., Bilyeu, G. P., Davis, D. L., Stephens, C. A., and Veal, G. R., "Geothermal Elastomeric Materials," Twelve-Months Progress Report, San Francisco Area Operations Office/1308-1, Sept. 1977.
[2] Ender, D. H., "Evaluation of Seal Materials for Deep Sour Gas Wells," Paper No. 102, presented at the International Corrosion Forum Devoted Exclusively to the Protection and Performance of Materials, Toronto, Ont., Canada, 14-18 April 1975.
[3] Mueller, W. A., Kalfayan, S. H., Reilly, W. W., and Inghan, J. D., "Development and Evaluation of Elastomeric Materials for Geothermal Applications," Department of Geothermal Energy/1026-1, Jet Propulsion Laboratory Publication 78-69, Sept. 1978.
[4] Schwartz, S., "Development of Improved Gaskets, Sealants and Cables for Use in Geothermal Well Logging Equipment," Hughes Aircraft Co., Report No. P78-31, Jan. 1978.
[5] Hawkins, W. L., *Polymer Stabilization*, Wiley-Interscience, New York, 1972, p. 6.
[6] Rosen, S. L., *Fundamental Principles of Polymeric Materials for Practicing Engineers*, Barnes and Noble, New York, 1971, p. 88.
[7] Hilado, C. J., *Pyrolysis of Polymers*, Vol. 13, *Fire and Flammability Series*, Technomic, Westport, Conn., 1976, p. 68.
[8] Kambe, H. and Garn, P. P., *Thermal Analysis: Comparative Studies on Materials*, Wiley, New York, 1974, p. 135.
[9] Roberts, J. D. and Caserio, M. C.., *Basic Principles of Organic Chemistry*, W. A. Benjamin, New York, 1965, p. 86.
[10] Pelosi, L. F. and Hackett, E. T., "Improved Steam Resistance for Fluoroelastomers," Paper presented at the 110th Meeting of the Rubber Div., American Chemical Society, San Francisco, Calif., 5-8 Oct. 1976.
[11] Gebhardt, I., Lengyel, B., and Torok, *Magyar Kemiai Folyoirat*, Vol. 67, 1962, p. 159.

L. E. Lorensen,[1] C. M. Walkup,[2] and C. O. Pruneda[3]

Polymeric and Composite Materials for Use in Systems Utilizing Hot, Flowing Geothermal Brine III

REFERENCE: Lorensen, L. E., Walkup, C. M., and Pruneda, C. O., "**Polymeric and Composite Materials for Use in Systems Utilizing Hot, Flowing Geothermal Brine III,**" *Geothermal Scaling and Corrosion, ASTM STP 717,* L. A. Casper and T. R. Pinchback, Eds., American Society for Testing and Materials, 1980, pp. 164–179.

ABSTRACT: This paper documents recent progress in our continuing program aimed at finding optimum polymeric materials for use in geothermal plants. Coatings and bulk specimens of high-performance polymers and composites were exposed to brine in the laboratory and in the field. Several levels of temperature and flow conditions were employed. Physical changes, scaling tendencies, and changes in surface characteristics resulting from exposure were studied by techniques including optical and scanning electron microscopy, energy-dispersive spectroscopy, and contact-angle measurements. Certain fluorocarbon and hydrocarbon polymers continue to respond favorably to the severe geothermal environment. Surface roughness, either present originally or developed during exposure, appears to be an important factor in promoting scaling and scale adhesion.

KEY WORDS: composite materials, brine, polymers, water resistance, thermal resistance, chemical attack, protective coatings, structural plastics, geothermal, chemical-resistant coatings, heat-resistant coatings, corrosion, scale

In developing alternate energy sources, severe material problems often accompany the engineering challenges. The development of geothermal energy illustrates this point; the geothermal resources containing the most energy are corrosive and erosive and often precipitate thick deposits of scale. These characteristics combine with high temperatures and pressures to create an inhospitable environment for most construction materials.

The geothermal fluids in question are rather unusual in that they are anaerobic or, usually, even reducing, because of their underground origin

---
[1] Chemist, Lawrence Livermore Laboratory, Livermore, Calif. 94550.
[2] Technical specialist, Lawrence Livermore Laboratory, Livermore, Calif. 94550.
[3] Chemist, Lawrence Livermore Laboratory, Livermore, Calif. 94550.

and the presence of hydrogen sulfide. This fact must be kept in mind when selecting candidate materials, because most polymers are normally evaluated for use in air. Materials with poor reputations because of oxidation sensitivity might perform well in the geothermal environment. Also, geothermal fluids can contain over 20 different elements and produce scale that cannot be duplicated in the laboratory. Therefore, a proper evaluation requires extensive field testing, in addition to the more easily controlled laboratory testing.

In 1973 the Lawrence Livermore Laboratory, Livermore, Calif., became involved in a program to study ways of utilizing the high-temperature, high-salinity brine prevalent underground in the Salton Sea area in Calif., [1].[4] Polymers and composites were studied to see whether commercially available or even experimental materials could solve the special set of problems involved [2,3,4]. The results were also expected to be of value in guiding any future custom syntheses of polymers for geothermal use. Other laboratories are also studying polymeric materials for geothermal applications [5-10].

In this progress report we describe our most recent results from exposing selected polymeric materials to a variety of field and laboratory conditions. Our objective has been twofold: to determine the material response in terms of both survival and scaling as a function of exposure conditions and to use various analytical techniques in looking for details that could explain why failure occurred and how scaling took place.

Experimental Procedure

Material specimens and actual parts were exposed to flowing brine and steam mixtures in different locations in two types of field pilot plant facilities: a direct conversion plant involving nozzle/turbine combinations and a multistage flashing plant. The specific exposure sites are described for each case in the order of the hottest to the coolest in Tables 1 and 2. A laboratory exposure test is described also.

Materials were exposed as either bulk or neat specimens or coatings on metal panels. With the latter, Teflon tetrafluoroethylene (TFE)[5] washers were used to insulate the metal panels from the mounting racks to which they were bolted. The racks with specimens attached were suspended in the hot brine in the test container or vessel. In the laboratory test, four specimens were place in a round-bottom flask. Each specimen leaned outward against the flask wall to allow full surface exposure to brine and vapor. The artificial brine was made by mixing ten of the most frequently found

---

[4]The italic numbers in brackets refer to the list of references appended to this paper.

[5]Reference to a company or product name does not imply approval or recommendation of the product by the University of California or the U.S. Department of Energy to the exclusion of others that may be suitable.

TABLE 1—*Exposure sites used in direct conversion field facility.*

Location	Specimen or Part Configuration	No. Tested	Conditions
Downhole	perforated tube or sheath	1	260°C, 2280 h, 550-m depth, occasional flow
Wellhead	specimens bolted to 2-m steel bar suspended in center of 76-mm pipe ("shish kebab" array)	37	190°C, 50 h (polymers), 20 h (Al and TFE), 1.6 to 1.9 MPa full well flow
Near wellhead	nozzle, 25 mm long, 12 mm throat diameter	3	190 to 60°C, 21 h, converging/diverging flow
Approximately 10 cm beyond nozzle	flat plates, 38 by 63 by 4 mm	9	approximately 60°C, 23 to 37 h, pH 3.7, 800 ft/s flow[a]
Approximately 30 cm beyond nozzle	10-cm steel pipe elbow	2	approximately 60°C, 23 to 37 h, pH 3.7, 800 ft/s flow[a]
Below pipe elbow	10-cm splash control tube	1	approximately 60°C, 23 to 37 h, pH 3.7, 800 ft/s flow[a]
Atmospheric splash tank or receiver	51 by 64 to 76 mm by 3 to 13 mm plates suspended in brine	23	approximately 60°C, 170 h, acidified upstream, pH 5.5

[a]ft/s × 0.3048 = m/s.

brine cations as chlorides in their natural molar ratio to a total concentration of 25 percent.[6]

After exposure, the specimens or parts were examined for condition and amount of scaling by ordinary weighing and measuring.[7] Then, in line with our second objective, many specimens were subjected to scale adhesion tests, optical and scanning electron microscopy including energy-dispersive spectroscopy, contact-angle measurements, and attenuated total reflectance infrared study. Standard techniques were used.

The scale adhesion tests require some comment. Aluminum cylinders 29 by 38 mm in length were bonded to the scaled surfaces by one of two methods. For thick scale, a specially formulated thixotropic epoxy was used. This approach was not suitable for thin scales because the adhesive could penetrate thin scale and bond to the subsurface; therefore, two layers of double-sided masking tape were used for thin scale. The strength of scale adhesion was measured by pulling apart the specimens in the Instron Universal Testing Machine, using a universal joint attachment to the

---

[6] The concentrations of salts in milligrams per litre were as follows: NaCl, 1.82 × 10^5; CaCl$_2$, 1.04 × 10^5; KCl, 4.24 × 10^4; MnCl$_2$·4H$_2$O, 1.48 × 10^4; SrCl$_2$·6H$_2$O, 1.81 × 10^3; LiCl, 1.73 × 10^3; ZnCl$_2$, 1.41 × 10^3; BaCl$_2$·2H$_2$O, 6.00 × 10^2; MgCl$_2$·6H$_2$O, 1.13 × 10^2; CuCl$_2$·2H$_2$O, 10.9.

[7] An analytical balance and machinists' micrometer were used (1) within 1 h for laboratory specimens and (2) after at least 3 days equilibration in the laboratory (68 ± 2°C, 50 ± 10 percent RH) for specimens returned from field tests.

TABLE 2—*Exposure sites, continued: multistage flash system and laboratory boiling brine test.*

Location	Specimen Configuration	No. Tested	Conditions
Second stage of multistage flash system	25 by 51 by 3 mm	13	150°C, prescaled, then 750 h, pH 4 to 6 (varied), steam/brine
Third stage of multistage flash system	25 by 51 by 3 mm	3	125°C, prescaled, then 750 h, pH 4 to 6 (varied), steam/brine
Fourth stage of multistage flash system	25 by 51 by 3 mm	10	105°C, prescaled, then 750 h, pH 4 to 6 (varied), steam/brine
Laboratory-partial immersion in hot brine	13 to 25 by 38 to 76 by 3 mm	19	106°C, 314 h, artificial brine, refluxing, under N_2

aluminum cylinder and a crosshead speed of 0.2 cm/min. With specimens in which the full 29-mm diameter circle of scale did not come off, the area that did debond was carefully measured with a planimeter and used in the calculations.

Results and Discussion

The experimental results form a rather extensive but incomplete matrix consisting of a variety of materials exposed to conditions varying in severity and then examined in several ways. We will discuss the results in terms of each examination method, noting the effect of exposure condition where appropriate.

Shape and Dimensional Changes

A Teflon TFE tube was used as a protective sheath for some specimens in a downhole test (Table 1); it survived with no difficulty. Had large differential pressures been involved, material flow or distortion would probably have been observed.

Three sets of specimens were exposed to flowing brine at the wellhead by bolting them down the length of a 19-mm square steel bar, shish-kebab fashion, and suspending the assembly in the center of a pipe. The first set was made up of ten aluminum specimens, and the second of nine Teflon TFE specimens. Both sets were used as controls to see if the mounting position or orientation had any effect on the results; none was found. The aluminum was very badly corroded; a highly voluminous fluff was present. The Teflon specimens, though lightly scaled, were unchanged; even the edges were still sharp and not eroded (Fig. 1). The third set of specimens included two Tefzel specimens, one with 30 percent glass and one with 30 percent carbon, and two polyphenylene sulfide specimens, one with 30 percent carbon and one with 40 percent glass; none of the four was changed. Present also were seven polysulfones or polyether sulfones with up to 40 percent glass reinforcement. All were melted and at least partially swept away, even though the reported heat distortion temperatures of these materials ranged from 174 to 216°C, temperatures which bracket the brine temperature of about 190°C. Since the Teflon TFE specimens with a heat distortion temperature of 55°C were unchanged in the same test, this property does not predict performance. The last seven specimens exposed in this third set were coatings of polyphenylene sulfide, Teflon perfluoroalkoxy (PFA), and Teflon TFE, with and without primers on steel and titanium. No physical damage occurred, except in the case of the Ryton polyphenylene sulfide (PPS)/Teflon TFE coating which had blistered.

Material in the nozzle configuration was subjected to the most severe combination of corrosive, erosive, scaling, pressure, and thermal stress of

FIG. 1—*Sections of shish-kebab assemblies after exposure: Teflon TFE (top), aluminum (bottom).*

any of our test specimens. Two Teflon PFA nozzles were exposed for 20 h. One nozzle in brine acidified to reduce scaling was unchanged except for an increase in throat diameter of 0.28 mm. Another exposed to unmodified brine gave the same result.

Steel panels coated with different fluorocarbons, polyphenylene sulfide and an epoxy/phenolic resin were subjected to the expanding brine in front of the nozzles to study corrosion prevention and scaling. They were physically undamaged except for the epoxy/phenolic resin, which had become checked.

Pipe elbows used to divert flow downward from the nozzles to an atmospheric splash tank (AST) were internally coated with Ryton PPS and primed Teflon fluorinated ethylene propylene (FEP), and tested for 40 h. Approximately 90 percent of the PPS coating debonded, but only 15 percent of the FEP debonded, and this in the area subjected to the most severe erosion. The need for stronger coating adhesion is clearly indicated.

A *Teflon TFE splash control tube* extending from the pipe elbow into the splash tank was unchanged except for scaling. The brittle scale could be removed easily by light hammering.

Twenty-three specimens were exposed in the AST. Acrylonitrile/butadiene/styrene (ABS) specimens containing 30 percent glass and a migrating internal lubricant were both badly swollen. An epoxy/phenolic coating was cracked, and a shock-cooled Teflon TFE specimen was warped; the other specimens were unchanged. The AST appeared to provide a very mild test environment.

In the multistage flash test facility, 26 specimens were exposed. No physical damage was observed except for warping with Kynar and rather severe surface degradation beneath the scale in the case of two polyester specimens.

Scaling Resistance of Different Materials

If a material can survive exposure, or be potentially useful up to a given level of exposure severity, its ability to resist scaling becomes important. Fortunately, in contrast to the effects of the other four geothermal stresses, scaling is most pronounced at lower temperatures, so that an effective antiscaling material need not have high thermal resistance.

In this study, scaling resistance was measured by weighing exposed specimens and calculating the weight change per square centimetre of surface area. In some cases the order of scaling resistance obtained this way was compared with that obtained by changes in thickness. This was done for verification purposes because scale could easily be lost by inadvertent mishandling of soft or brittle materials. The agreement between the two methods was reasonable.

The quantity of scale accumulated per square centimetre was measured for the shish-kebab, flash tank, and AST test specimens (Table 3).

Since the test temperatures were 190, 150 and 105, and 60°C, and the times 50, 750, and 170 h, respectively, and it is not certain that the scaling rate is linear with time; the results must be viewed independently. The quantities of scale on the shish-kebab specimens ranged from 4 to 42 mg/cm^2; on the specimens from the three flash tanks, from 1 to 139 mg/cm^2; and on the specimens from the AST, from 0.1 to 8 mg/cm^2. The very low values observed in the case of the AST specimens are surprising, although other experimenters reported that the 125°C third stage produced more severe scaling than did the cooler fourth stage in acidified flash tank tests [11]. In contrast to other upstream specimens, those in the AST were submerged beneath essentially single-phase brine.

Because the materials selected for testing had different upper temperature limits, and because of limitations on the available test time and space, not every material could be tested in every location. However, rather clear trends in material response are evident. Fluorocarbon and hydrocarbon polymers, which have low surface energy, have the least scale. Next come those with more polar surfaces, such as polyesters and polysulfide. Polymers reinforced with glass or carbon fibers have the most scale. Phenolic resin seems to be a poor binder as regards scaling Since differences in microstructure or crystallinity might affect scaling, we tested both injection-molded and compression-molded Teflon PFA, and normal and shock-chilled Teflon TFE. No effects were noted.

Scale Adhesion Results

In an effort to learn more about the mechanism of material failure and scaling, studies were directed toward the interface, since it is so intimately involved.

The scale adhesion test was done on 7 tape-bonded and 22 epoxy-bonded specimens (see Experimental Procedure). The data are shown in Table 4. Despite some distressing scatter in several instances, the data are worth comment. The tests on thin scale required the use of double-sided masking tape to bond the aluminum cylinder to the scale. Very low adhesion values were observed. Although thin scale may be bonded to surfaces less firmly than thick scale, the scale tested might have been moved slightly by the 2-kg weight used to ensure good contact between the scale and the tape adhesive. The higher values observed when Teflon TFE scale was bonded with epoxy, as opposed to tape, support this. The use of the epoxy for bonding to scale gave a far greater range of values, again, with much thicker scale. The control values on unexposed material showed that the

172 GEOTHERMAL SCALING AND CORROSION

TABLE 3—*Quantity of scale gathered by different materials, mg/cm^2.*[a]

Shish-Kebab Specimens, 190°C		Flash Tank (Second Stage) Specimens, 150°C		Flash Tank (Fourth Stage) Specimens, 105°C[b]		Atmosphere Splash Tank Specimens, 60°C	
Teflon PFA[c], on steel	4.4	Dienite X-555	1.3	Dienite X-555	4.1	Teflon PFA, extruded	0.3
Teflon TFE	6.1	Teflon PFA, extruded	1.8			Teflon PFA, injection molded	0.4
Teflon PFA, on Ti	8.5	Teflon PFA, injection molded	2.6			polypropylene	0.5
				epoxy A	5.1	Kynar	0.7
Ryton PR01, on Ti	11	Kynar[e] (in brine phase)	2.7	epoxy B	6.6	polypropylene/MILTP[i]	1.0
Ryton P-3, on steel	15						
Teflon FEP[d], on Ti	16	Teflon FEP	3.1			Teflon FEP	0.1
Ryton/TFE, on steel	18	Kynar (in steam phase)	4.2			Tefzel	1.3
						polycarbonate	1.5
						polycarbonate	1.6
						polyphenylene oxide	2.1
		Dienite X-559, gl	9.2	Dienite X-559, gl	30	polycarbonate/MILTP	2.2
		polymide K-601, gl	14	polymide K-601, gl	67	polypropylene, 30 gl	2.3
		Dienite X-557, gl	25	Dienite X-557, gl	48	polyvinyl chloride 15 gl	2.4
		Dienite X-558, gl	34	Dienite X-558, gl	71	Teflon TFE, shock cooled	2.5
		phenolic, silica	36	phenolic, silica	139	Teflon TFE, untreated	2.7
Tefzel, 30 gl[g]	19	phenolic, gl	37	phenolic, gl	50	Tefzel, 30 gl	2.9
Ryton P-3, on Ti	20	phenolic, car	59	phenolic, car	126	polyphenylene sulfide, 40 gl	4.2
Polyphenylene sulfide, 30 car[h]	22					ABS[j]/MILTP	5.3
Polyphenylene sulfide, 40 gl	26						
Tefzel, 30 car	42					ABS, 30 gl	8.4

[a] Quantities increase down the columns, except where horizontal comparisons were possible.
[b] Third stage contained only Teflon FEP with 11 mg/cm², and two polyesters with degraded surfaces.
[c] Teflon perfluoroalkoxy.
[d] Fluorinated ethylene-propylene.
[e] Polyvinylidene fluoride.
[f] Polyacrylonitrile/butadiene/styrene.
[g] Thirty percent glass (gl) fiber reinforcement.
[h] Thirty percent carbon (car) fiber reinforcement.
[i] MILTP = Migratory internally lubricated thermoplastic. It was hoped that migration of the internal lubricant to the surface would reduce scale adhesion.

TABLE 4—*Strength of scale adhesion to polymers.*

		Tape[a]	Epoxy[a]	
Polymer Substrate	Exposure Site[h]	Exposed, kg/cm^2	Exposed, kg/cm^2	Unexposed, kg/cm^2
Polypropylene/MILTP[b]	AST	0.1		
Dienite 557, gl[c]	FT		0.4	
Teflon TFE[d]	SK/FT	0.5 average (0.4, 0.4, 0.1, 1.1, 0.6)[g]		
Tefzel[e]	AST	0.7		
Dienite 558, gl	FT		0.8	
Ryton[f]/TFE on steel	SK		2.0	0.9 average (0.7, 1.1)
Ryton P-3 on steel, Ti	SK		2.6 average (0.1, 5.2)	1.1
Polyester	AST		2.8	1.1
Tefzel, 30 carbon	SK		4.6	
Teflon TFE	SK		4.8 average (2.1, 4.5, 7.8)	
Epoxy	AST		6.6	
Polyphenylene sulfide, 30 gl	SK		8.3 average (7.7, 8.8)	
Polyphenylene sulfide, 40 gl	SK, AST		9.4 average (6.5, 11.0, AST 10.8)	
Tefzel, 30 gl	SK		11.4 average (8.8, 14)	

[a] Method used to attach Instron fixture to top of scale.
[b] Migrating internal lubricant.
[c] Glass (gl) reinforced poly-1,2-butadiene.
[d] Polytetrafluoroethylene.
[e] Polyethylene-tetrafluoroethylene.
[f] Polyphenylene sulfide.
[g] Average of 0.5, from individual values noted.
[h] AST = atmospheric splash tank; FT = flash tank; SK = shish-kebab.

epoxy, at full coverage, would have made only a minor contribution had it soaked completely through the scale; thus the scale-to-substrate bonding is rather strong. The data suggest several things: the exposure site was not of first-order importance; the internal migrating lubricant in polypropylene may be working as hoped; the Dienites have good scale-release properties; and glass present as reinforcement can be detrimental, particularly in a machined surface. The Dienites contained glass but were tested as received, which means that the glass was buried under a gel coat with a mold-release agent probably present.

Optical Microscopy

Unexposed surfaces, as well as those from which scale had been pulled during adhesion tests, were examined with a stereo zoom microscope with up to ×140 magnification. Among the significant observations was that Teflon TFE even before exposure has a relatively bumpy, rutted, or pitted surface, probably related to the void content normally present as a result of the sintering process used in its manufacture. Islands of scale when moved slightly appeared to be elastically attached to the polymer by a root structure. Ryton became rougher during exposure. Glass fibers, exposed at the ends of reinforced specimens were not scaled, which suggests that they contribute to scaling simply by increasing roughness. Machine marks were seen often. In Tefzel the tops of ridges were feathered and after exposure were observed to be a locus of scale accumulation. In general these observations were consistent with those obtained by scanning electron microscopy.

Scanning Electron Microscopy and Energy Dispersive Spectroscopy

Twenty-four specimens, approximately one third of those exposed, were examined using scanning electron microscopy (SEM). In some cases, the underside of the scale, as well as the surfaces from which scale had been removed during adhesion tests, was examined. It was noted that surface disturbance increased with the severity of exposure, as expected. The observed roughening appeared to contribute to the tendency to collect scale and arose from several sources. Machining left feathered edges which collected and held scale (Fig. 2). Fiber presence contributed to scaling, particularly where the specimen was machined. Machining probably exposes fiber ends and tears off and drags stiff fragments, leaving a rough surface (Fig. 3). Although neither glass nor carbon fibers nucleated scale, glass fibers apparently contributed to very strong scale adhesion in at least one case. Sharp peaks perpendicular to the surface covered the area. We suspect that roughening and consequent scaling might arise from inherent microstructure defects present because of incomplete knitting of particles used in the molding or compaction process. For example, granular areas were noted in the unexposed surface of Teflon TFE (Fig. 4).

Scale formation on the low-surface-energy polymer Teflon TFE is a very interesting phenomenon. The formation of fibrils or surface fibers noted in an earlier report [3] was found in several specimens that had been exposed to brine at 150°C or hotter (Fig. 5). Teflon exposed at lower temperatures had no fibers. While microscopically observing scale platelets on Teflon, we moved them slightly with a prod and found them to be elastically attached to the substrate polymer. The underside of the scale had fibers imbedded in it as well as a very unusual cellular structure (Fig. 6). Thus the development of fibers during exposure and the presence of voids and granules in the surface, resulting from manufacturing procedures, account for the scaling observed with Teflon TFE. It has been reported that fluoro-

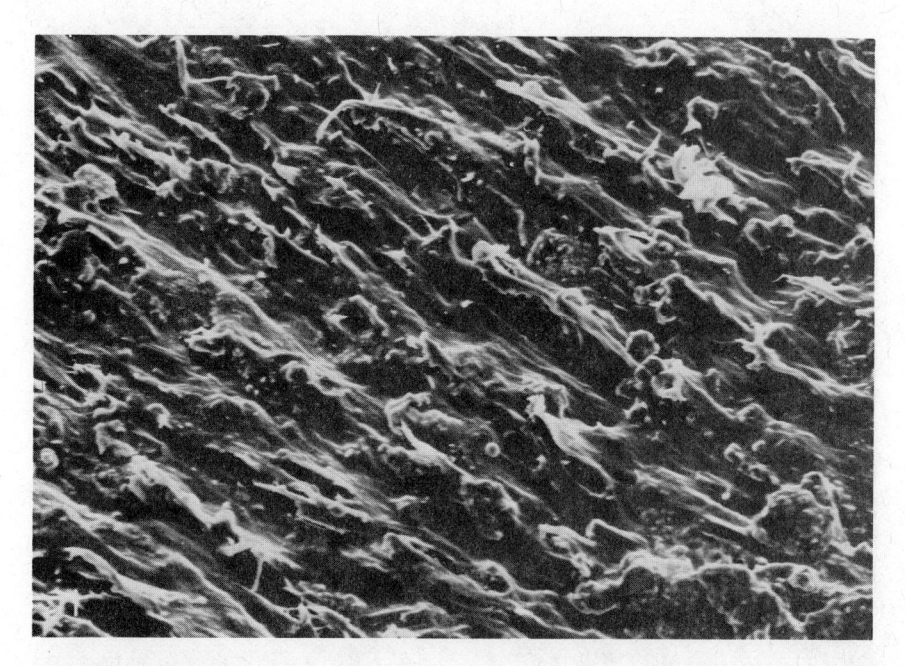

FIG. 2—*Scale collected on feathered edges of machining marks on Tefzel/30 carbon. (Scale mark indicates 20 μm).*

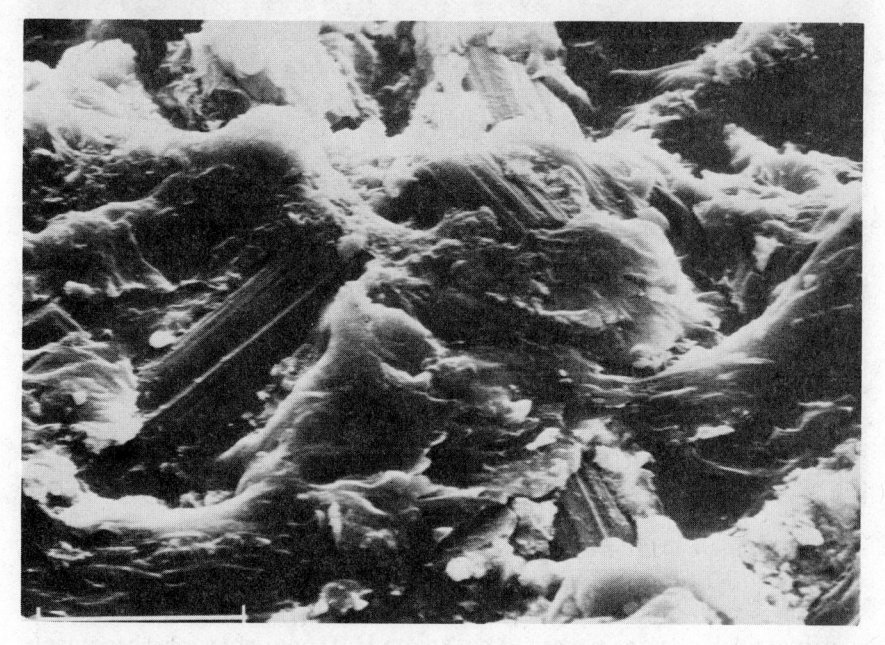

FIG. 3—*Fiber damage and roughness on exposed polyphenylene sulfide/30 carbon. (Scale mark indicates 10 μm).*

176 GEOTHERMAL SCALING AND CORROSION

FIG. 4—*Granular areas on unexposed Teflon TFE surface. (Scale mark indicates 20 µm).*

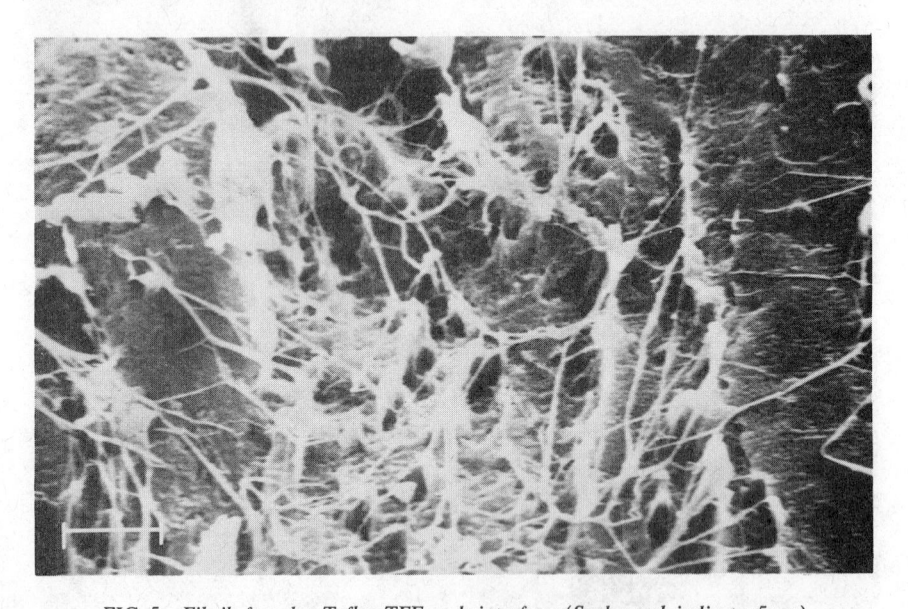

FIG. 5—*Fibrils found at Teflon TFE-scale interface. (Scale mark indicates 5 µm).*

polymers can readily transfer to adjacent surfaces [*12*] and also that certain colloidal oxides can be deposited on Teflon [*13*], both of which phenomena appear to be related to our observations on scaling.

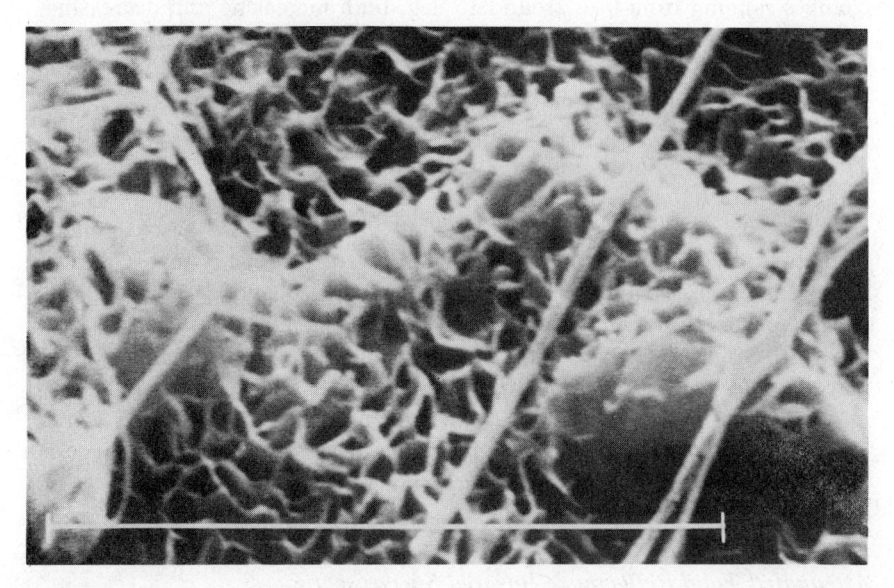

FIG. 6—*Fibrils and cellular structure on underside of scale from Teflon TFE. (Scale mark indicates 5 μm).*

Energy-dispersive spectroscopy revealed that scale residues are composed of the expected variety of elements ranging from magnesium to iron and lead.[8] Two-micrometre spheres of calcium chloride were noted. No obvious correlation between scale composition and specimen type or location was found.

Contact-Angle Measurements on Exposed and Unexposed Specimens

It is well known that contact angles can reveal considerable information about surface character [14]. Increasing roughness for a given surface causes contact angles to increase by as much as 50 deg. Increasingly polar surfaces cause contact angles with polar fluids to decrease. Therefore, such measurements were made as a potentially useful and simple tool to augment microscopy in measuring physical and chemical changes, along with scale deposition. As an illustration, Teflon FEP was exposed to hot, flowing brine for 23 h. The protected areas showed contact angles with water of 102 to 105 deg, as did the scale-free exposed areas, whereas the lightly scaled areas showed contact angles in the range of 40 to 60 deg.

The areas from which scale had been removed in the adhesion tests were measured for changes in the contact angle of water. Field exposure caused

---

[8] Energy dispersive spectroscopy (EDS) is useful only for elements above neon.

changes ranging from 0 to around 15 deg, both increasing and decreasing. The results were relatively consistent with the surface characteristics revealed by SEM. Roughening markedly affected the contact angle, even tolerating small amounts of isolated scale in its positive effect. As scale increased, however, it could become dominant and lower the contact angle. Thus far we have been unable to dissolve residual scale and therefore cannot separate the effects.

The specimens from the 106°C boiling brine test were also examined with contact-angle measurements. These were sulfone, vinyl chloride, phenylene oxide, carbonate, sulfide, fluorocarbon, phenolic, polyester, and epoxy-type polymers, many with fiber reinforcement. Because of the very mild test conditions, the changes in weight and dimension (bulk) were 1 percent or less, which suggests that whatever changes occurred were located primarily at the specimen surfaces. In almost all cases, the contact angles first decreased and then increased during the 314-h test. This suggests competing processes of changing importance, but the results are not yet understood.

Attenuated Total Reflectance Infrared Spectroscopy

Infrared spectroscopy in the attenuated total reflectance (ATR) mode has been used to detect changes in surface chemistry. It was applied to a polyester and to polyphenylene sulfide to see if changes during exposure could be detected. No differences were noted.

Summary and Conclusions

In our study of material response to geothermal environments, we have screened another set of commercial high-performance polymers, many of which are fiber-containing composites. The earlier findings that aromatic, fluorocarbon, and cross-linked structures responded satisfactorily are still supported, although polysulfone materials were badly degraded.

Regarding scaling, in general the low-surface-energy polymers scaled least, and scale adhered less strongly to them. Surface roughness, from whatever source, proved to be an important and potentially overriding factor in promoting scaling. Beyond simple physical entrapment of particles from solution, which nucleate growth, there are probably other phenomena at work. An example might be a concentration of frictionally generated (triboelectric) charges gathered on hairs, feathered edges, or other discontinuities, which could attract charged particles from solution, thus promoting growth. Another factor important in the scaling process might be the presence of a two-phase flow in which surface areas are exposed to a continual alternation of liquid and vapor, which could cause sudden, very small changes in localized concentrations of key scale-promoting compounds.

Acknowledgments

We wish to thank G. E. Tardiff and R. D. McCright for assistance in performing field tests, W. J. Steele for performing scanning electron microscopy and energy-dispersive spectroscopy, and R. H. Sanborn for ATR-infrared scans. The work was performed under the auspices of the U.S. Department of Energy by the Lawrence Livermore Laboratory under Contract No. W-7405-ENG-48.

References

[1] Austin, A. L., Higgins, G. H., and Howard, J. H., "The Total Flow Concept for Recovery of Energy from Geothermal Hot Brine Deposits," UCRL-51366, Lawrence Livermore Laboratory, Livermore, Calif. 1973.
[2] Lorensen, L. E., "Materials Screening Program for the LLL Geothermal Project," UCID-16513, Lawrence Livermore Laboratory, Livermore, Calif., 29 May 1974. (This paper also appears in the *Proceedings* of the 14th Annual American Society of Mechanical Engineers, Symposium, Albuquerque, N. Mex., Feb. 1974).
[3] Lorensen, L. E., Walkup, C. M., and Mones, E. T., *Proceedings*, Vol. 3, Second United Nations Symposium on the Development and Use of Geothermal Resources, San Francisco, Calif., 1975, pp. 1725-1731.
[4] Lorensen, L. E. and Walkup, C. M., "Polymeric and Composite Materials for Use in Systems Utilizing Hot, Flowing Geothermal Brine II," Preprint UCRL-81019, presented at the American Institute of Chemical Engineers Meeting, Philadelphia, Pa., June 1978, Lawrence Livermore Laboratory, Livermore, Calif., 1978.
[5] Schwartz, S. and Basiulis, D., "Development of Improved Gaskets, Sealants and Cables for Use in Geothermal Well Logging Equipment," Report No. P78-31, Hughes Aircraft Co., Culver City, Calif., Jan. 1978.
[6] Steinberg, M., Kukacka, L. E., Fontana, J., Zeldin, A., Amaro, J., Sugama, T., Carciello, N., Reams, W., Auskern, A., Horn, W., Colombo, P., Romano, A., DePuy, G., W., Causey, F. E., Cowan, W. C., Lockman, W. T., and Smoak, W. G., "Alternate Materials of Construction for Geothermal Applications," Series of Progress Reports, Brookhaven National Laboratory, Upton, N.Y., 1973 to present.
[7] Moiseev, Y. V., Markin, V. S., and Zaikov, G. E., *Russian Chemical Reviews*, Vol. 45, No. 3, 1976, pp. 246-266, in translated journal.
[8] Polymeric Materials Seminar and Workshop Proceedings, Task Group on Geothermal Seals, ASTM Subcommittee D11.36 on Seals, Boston, Mass., 27 June 1978.
[9] Cassidy, P. E. and Rolls, G. C., "Polymeric Materials in Geothermal Energy Recovery," Department of Energy Materials Science Workshop on Polymers, Case Western Reserve University, Cleveland, Ohio, 27-29 June 1978.
[10] Workshop/Symposium on Materials in Geothermal Energy Systems, Department of Energy Division of Geothermal Energy and Radian Corp., Lakeway Inn, Austin, Tex., 23-25 May 1978.
[11] Harrar, J. E., Otto, C. H., Jr., Deutscher, S. B., Ryon, R. W., Tardiff, G. E., "Studies of Brine Chemistry, Precipitation of Solids and Scale Formation at the Salton Sea Geothermal Field," Report No. UCRL-52640, Lawrence Livermore Laboratory, Livermore, Calif., 1979.
[12] Dwight, D. W., "Surface Energy and Adhesion in Fluoropolymers," American Chemical Society Colloid and Surface Chemistry Division, New York, N.Y., 5-8 April 1976.
[13] Kenney, J. T., Townsend, W. P., and Emerson, J. A., *Journal of Colloid and Interface Science*, Vol. 42, No. 3, March 1973.
[14] Dettre, R. H. and Johnson, R. E., "Contact Angle, Wettability and Adhesion," *Advances in Chemistry Series*, No. 43, American Chemical Society, Washington, D.C., 1964, Chapter 8, pp. 136-144.

J. J. Fontana,[1] and A. N. Zeldin[2]

Concrete Polymer Materials as Alternative Materials of Construction for Geothermal Applications—Field Test Evaluations

REFERENCE: Fontana, J. J. and Zeldin, A. N., "**Concrete Polymer Materials as Alternative Materials of Construction for Geothermal Applications—Field Test Evaluations,**" *Geothermal Scaling and Corrosion, ASTM STP 717,* L. A. Casper and T. R. Pinchback, Eds., American Society for Testing and Materials, 1980, pp. 180-193.

ABSTRACT: A serious problem in the development of geothermal energy is the availability of durable and economical materials of construction for handling hot brine and steam. Hot brine and other aerated geothermal fluids are highly corrosive, and they attack most conventional materials of construction. The Brookhaven National Laboratory, Upton, N.Y., has been investigating the use of concrete polymer materials as alternative materials of construction for geothermal processes. To date, successful field tests have been demonstrated at the Geysers, Calif., the U.S. Bureau of Mines Corrosion Facility, Niland, Calif., and the Department of Energy Geothermal Facility at East Mesa, Calif. This report is a survey of field and laboratory evaluations of concrete polymer materials that have been shown to be durable and economical as alternative materials of construction.

KEY WORDS: geothermal, polymer concrete, polymer-impregnated concrete, brine, aggregate, monomer, liner, corrosion, scaling

A serious problem in the development of geothermal energy is the lack of durable and economical materials of construction for handling hot brine and steam. Hot brine and other aerated geothermal fluids are highly

[1] Research scientist, Brookhaven National Laboratory, Process Sciences Division, Department of Energy and Environment, Upton, N.Y. 11973.
[2] Associate chemical engineer, Brookhaven National Laboratory, Process Sciences Division, Department of Energy and Environment, Upton, N.Y. 11973.

corrosive, and they chemically attack most conventional materials of construction. Corrosion and scale encrustations have been encountered in all geothermal plants and, to various degrees, adversely affect plant lifespans and power output. To date, carbon steel has been the primary material of construction, but the use of expensive materials such as Type 316 stainless steel and titanium-base alloys may be required to ensure long-term operation.

Concrete polymer materials have been under development at Brookhaven National Laboratory (BNL), Upton, N.Y., for some time [1-3].[3] Concrete polymer materials can be divided into two categories: (1) polymer-impregnated concrete and (2) polymer concrete. Both materials have the desired properties of high strength and long-term durability.

Polymer-impregnated concrete is obtained as follows. Precast concrete that has been cured is dried to remove the free water (approximately 3 weight percent). This can be done in an air-circulating oven at 100 to 150°C. The concrete is then cooled and placed in a closed vessel evacuated to a few millimetres of mercury absolute. Liquid-monomer–containing initiator is introduced, and, to ensure rapid and thorough permeation of the concrete, an overpressure of air or nitrogen up to 689 kPa is applied. The rate of permeation depends on the viscosity of the monomer and the overpressure in the vessel. The excess monomer is drained off, and the curing of the monomer *in situ* is accomplished with steam or hot water directly in the vessel.

Polymer concrete is defined as a concrete in which the aggregate is bound in a dense matrix with a polymer binder. Well-graded aggregate is mixed with monomer systems containing initiators and promoters and placed in forms just like those used for casting portland cement concrete. The curing can be obtained at ambient or elevated temperatures. Sometimes combinations of ambient temperature curing with final curing at elevated temperatures up to 150°C are used.

The work has progressed to the point that the feasibility of several energy-related applications has become apparent [4]. These include the following:

(*a*) polymer-impregnated concrete, condensate and cooling water pipes, distribution piping, and prestressed concrete reactors for desalination plants; and

(*b*) polymer concrete, lining of pipes and vessels to handle geothermal brines and acid solutions, well-cementing material for oil and geothermal wells, and pipes for condensate and cooling water as well as distribution piping.

---

[3] The italic numbers in brackets refer to the list of references appended to this paper.

Field Test Evaluations

Office of Saline Water Desalting Facility, Freeport, Tex.

Partial impregnation and coating techniques were first tested in a field experiment at the Office of Saline Water (OSW) Desalting Facility at Freeport, Tex., in August 1972. The field application required the portland cement liner of a vertical tube evaporator (VTE) to be partially impregnated to a depth of approximately 6 mm. The VTE was fabricated of steel plate lined with 2.5 cm of lumnite cement mortar with steel mesh reinforcing. The combination of salt water, temperatures elevated to 177°C, and the acid washing of the vessels caused numerous maintenance problems. Brookhaven National Laboratory was asked by OSW to apply a protective sealing on the mortar surfaces.

A monomer mixture of 60 weight percent styrene/40 weight percent trimethylolpropane trimethacrylate (TMPTMA) had been used as an impregnating material for some time at BNL [2]. Polymer-impregnated specimens were tested in autoclaves for up to one year with little or no deterioration evident. Figure 1 shows the thermograms of polymer before and after a 239-day exposure to brine at 143°C. There is no apparent change in the curves, which indicates that the polymer was not degrading. Figure 2 shows the compressive strengths of polymer-impregnated concrete as a function of exposure to brine at 143°C. A decrease in strength is noted with exposure time. The compressive strength of the specimens after exposure to brine for 300 days was 69 MPa compared with an initial compressive strength of 138 MPa.

The monomer mixture had a viscosity of approximately 0.001 Pa·s and could be applied on the mortar surface with paintbrushes or rollers to be absorbed into the pores of the mortar. The surfaces were dried with infrared bulbs, as seen in Fig. 3.

The monomer was applied several times and was readily absorbed into the mortar. After the soak coat had been applied, a seal coat with a viscosity of approximately 0.4 Pa·s was applied. The viscosity of the seal coat was obtained by dissolving a polyester resin in the soak coat monomer system. The seal coat prevented the evaporation of the soak coat before it had been cured and also filled and sealed cracks and other flaws in the mortar surface.

In addition to the VTE, mortar-lined pipes and elbows 15, 20, and 25 cm in diameter were also treated with polyurethane foam swabs saturated with monomer. The saturated swabs were passed back and forth through the pipes until the mortar liner was saturated with monomer. A typical operation is shown in Fig. 4.

The mortar liner from several of the multiflash distillation chambers was disintegrated. The liner on the floor of two chambers and a vertical wall

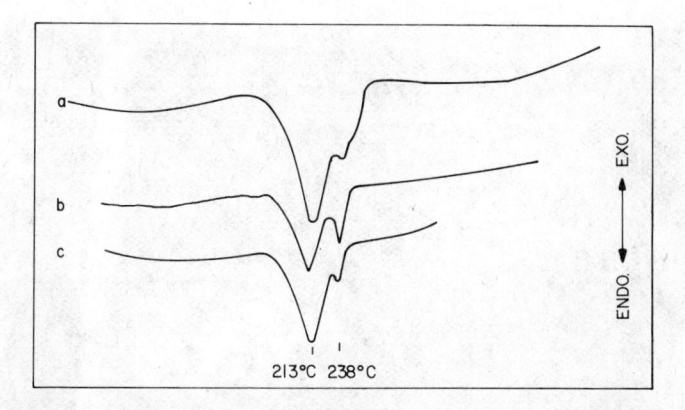

FIG. 1—*Differential thermal analysis of 60 weight percent styrene/40 weight percent trimethylolpropane trimethacrylate after various exposures:* (a) *bulk polymer, as prepared;* (b) *bulk polymer, 48 h at 143°C; and* (c) *from polymer-impregnated concrete, 5740 h at 143°C.*

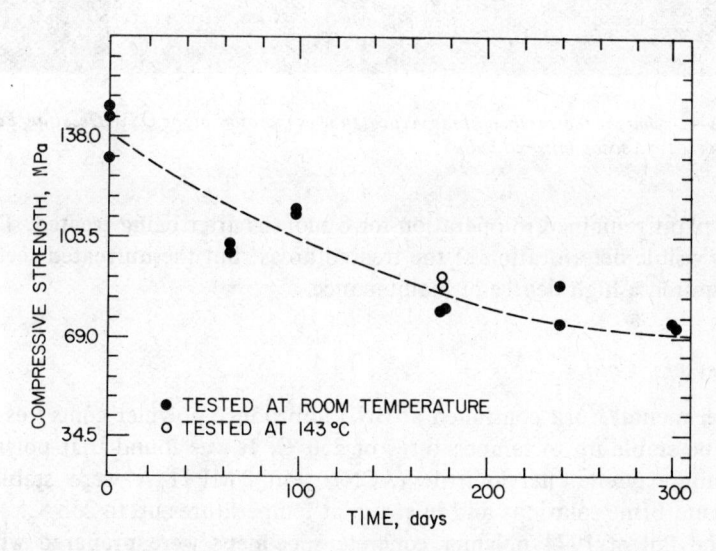

FIG. 2—*Compressive strength of polymer-impregnated concrete containing 60 weight percent styrene/40 weight percent trimethylolpropane trimethacrylate after immersion in brine at 143°C.*

of a third chamber were completely removed and replaced with a polymer concrete liner. The polymer concrete mortar was made by using graded silica sand with a 60 weight percent styrene/40 weight percent TMPTMA monomer system containing initiator and promoter. The mixed mortar was placed in the cavities, finished, and allowed to cure at ambient temperatures.

FIG. 3—*Drying the mortar liner of the vertical tube evaporator, at the OSW Desalting Facility in Freeport, Tex., using infrared bulbs.*

The plant remained in operation for 8 months after being treated. There was no visible deterioration of the treated areas, but the untreated sections still required a high degree of maintenance.

The Geysers, Calif.

Experimental work continued at BNL to produce polymer concretes that would be stable up to temperatures of 250°C. It was found that polymers containing styrene, acrylonitrile (ACN), and TMPTMA were stable in 25 percent brine solutions and in steam at temperatures up to 238°C.

In the fall of 1974 polymer concrete specimens were prepared with a monomer system containing 50 weight percent styrene/33 weight percent ACN/17 weight percent TMPTMA and three different aggregate systems. Two of the aggregate systems contained coarse stone (crushed limestone or river gravel) and silica sand. The third system contained silica sand and portland cement. The specimens were exposed to flowing dry steam at 238°C at the Geysers [5] and returned after exposures of 30, 60, and 90 days. As can be seen in Fig. 5, only one group of specimens did not deteriorate. This group contained silica sand and cement. Similar specimens tested in autoclaves simultaneously with the field experiments gave the

FIG. 4—*Typical partial impregnation of mortar-lined steel pipes, using polyurethane foam saturated with monomer.*

same results. Recent experimental work at BNL indicates that a bond exists between the vinyl-type monomers and the cement phase [6]; thus cement-containing specimens would be expected to have higher thermal stability and long-term durability.

The compressive strengths of polymer concrete specimens tested at the Geysers in flowing dry steam at 238°C show a decay for the first 30 to 60 days and then remain constant for periods of one year or more. Table 1 shows some of the results obtained. These data have been duplicated in the laboratory in autoclave tests.

The early decays in strength are attributed to unreacted monomer in the polymer concrete. Thermogravimetric analysis has confirmed that 0.1 to 0.5 weight percent of styrene monomer is unreacted. In order to achieve a higher degree of polymerization, a combination of two initiators is used [7]. One initiator has a low activation temperature of approximately 60°C, and the second has an activation temperature of approximately 140°C. Polymer concrete specimens with two initiators are cured at two temperature ranges (16 h at 60°C plus 4 h at 150°C). This type of curing produces specimens with higher original strengths and lower strength decays after exposure to geothermal environments.

FIG. 5—Polymer concrete specimens returned from the Geysers after exposure to flowing dry steam at 238°C.

TABLE 1—*Polymer concrete specimens exposed to flowing dry steam at 238°C at the Geysers.*[a]

Exposure Time, days	Monomer System, weight percent	Compressive Strengths, MPa
0	50 S[b]/35 ACN[c]/17 TMPTMA[d]	75
0	55 S/36 ACN/9 TMPTMA	75
60	50 S/35 ACN/17 TMPTMA	31
90	50 S/35 ACN/17 TMPTMA	30
90	55 S/36 ACN/9 TMPTMA	34
360	50 S/35 ACN/17 TMPTMA	30
360	55 S/36 ACN/9 TMPTMA	30

[a] All polymer concrete specimens contained an aggregate system composed of 90 weight percent silica sand/10 weight percent portland cement.
[b] S = styrene monomer.
[c] ACN = acrylonitrile monomer.
[d] TMPTMA = trimethylolpropane trimethacrylate cross-linking agent.

U.S. Bureau of Mines Corrosion Facility, Niland, Calif.

A polymer concrete-lined pipe was installed at the U.S. Bureau of Mines Corrosion Facility at Niland, Calif., in the Salton Sea geothermal field. This is considered to be the most aggressive corrosive fluid known to date. It contains >250 000 ppm brine at wellhead temperatures of 240 to 260°C. The lined pipe was installed in the inlet line of the corrosion facility and was in service for 3 months. Figure 6 shows the pipe installed. It was 1.2 m long and had an inside diameter of 7.6 cm. The liner was made of polymer concrete, containing 55 weight percent styrene/36 weight percent ACN/9 weight percent TMPTMA. The aggregate system used was 70 weight percent silica sand/30 weight percent Type III portland cement.

When the pipe was returned it was sectioned and examined. A scale buildup of 0.08 to 0.16 cm was evident on the surface of the polymer concrete liner. Three 15.2-cm sections were tested for shear bond strength at the interface between the liner and the steel pipe. Figure 7 shows the scale buildup and the polymer concrete liner after it had been pushed out of the steel pipe. Evidence of the bond between the steel pipe and the polymer concrete can also be seen. Table 2 gives the shear bond strengths of the liner at various temperatures. All indications to date substantiate the fact that polymer concrete liners bond well to steel pipes, and scale buildup appears to be slightly less than that reported in untreated steel pipes.

Department of Energy Geothermal Facility, East Mesa, Calif.

Two 2.4 m long by 5.0 cm in inside diameter polymer concrete-lined pipes

FIG. 6—*Polymer concrete-lined steel pipe installed at the U.S. Bureau of Mines Corrosion Facility at Niland, Calif., in the Salton Sea geothermal field.*

were installed at the East Mesa Geothermal Facility for one year. A 1.6-cm diameter orifice was installed between the two pipes to induce flashing of the brine fluid. After operation for one year, the scale buildup was less than 2.0 mm. This can be seen in Fig. 8. The personnel who operate and maintain the facility indicated that this result was extremely good. In many cases scale has to be cleaned out every 3 to 6 months from the steel pipe that is normally used.

A polymer concrete-lined steam separator and associated piping were constructed at BNL and installed at the East Mesa Geothermal Facility in January 1979. As of this date, information concerning scale buildup or corrosion durability has not been available. Figure 9 shows the installation of the steam separator and associated piping.

Summary and Conclusions

Concrete polymer materials as alternative materials of construction have been tested in the field and in the laboratory. In all cases, results obtained in the laboratory have been confirmed in the field.

To date, the feasibility of using polymer concrete liners for steel pipe or

FIG. 7—*A section of polymer concrete pipe after 3 months of service at the U.S. Bureau of Mines Corrosion Facility at Niland, Calif.*

vessels to extend their service life in geothermal environments has been confirmed. Aggressive corrosive fluids do not deteriorate polymer concretes at temperatures exceeding 200°C. However, it is important that the correct polymer concrete composition be chosen for the location in which it is to be used.

Preliminary results from economic evaluations [4] indicate that concrete polymer materials are cost effective in geothermal processes. Large cost benefits can be accrued by the use of polymer concrete-lined carbon steel components as replacements for stainless steel, tantalum, and Hastelloy alloy in condensate piping systems, reinjection lines, and steam separators. Uses of polymer concrete or polymer-impregnated concrete in cooling towers and district heating systems and to protect concrete surfaces also appear cost effective. Table 3 gives a cost comparison of concrete polymer materials for geothermal application.

Acknowledgment

This work was performed under the auspices of the U.S. Department of Energy under Contract No. EY-76-C-02-0016.

TABLE 2—*Bond strengths in shear at interface between polymer concrete and steel.*[a]

Test Temperature, deg C	Shear Strength, MPa
23	10.26
100	10.46
200	8.89

[a] Polymer concrete manufactured with the following monomer and aggregate system: monomer = 55 weight percent styrene/36 weight percent acrylonitrile/9 weight percent trimethylolpropane trimethacrylate, aggregate = 70 weight percent silica sand/30 weight percent Type III portland cement.

FIG. 8—*Polymer concrete–lined steel pipe installed at the East Mesa Geothermal Facility, Calif., after one year.*

FIG. 9—*Polymer concrete-lined steam separator installed at the East Mesa Geothermal Facility, Calif.*

192 GEOTHERMAL SCALING AND CORROSION

TABLE 3—Cost comparison for use of concrete polymer materials for geothermal applications.*

Use	Current Material	Estimated Cost, $/unit	Type of Concrete Polymer Material	Estimated Cost, $/unit	Benefit, $/unit
The Geysers					
Condensate and cooling water pipe	stainless steel	120/ft[a]	carbon steel with PC liner	30/ft[a]	90/ft
Cooling towers (all wetted parts)	stainless steel or aluminum	1.15/lb[b] 1.15/lb[b]	PC PC on carbon steel	0.10/lb 0.30/lb[b]	1.05/lb .85/lb
Protective liners on concrete	coal tar epoxies	1.20/lb	PC	0.10/lb	1.10/lb
Imperial valley brines					
Steam separators	stainless steel Hastelloy alloy tantalum	1.25/lb[c] 8.56/lb[c] 120/lb[c]	PC on carbon steel PC on carbon steel PC on carbon steel	0.40/lb[c] 0.40/lb[c] 0.40/lb[c]	.85/lb 8.16/lb 119/lb
Reinjection lines	stainless steel	120/ft[a]	carbon steel with PC liner	30/ft[a]	90/ft
Acid-handling system	stainless steel	26/ft[d]		5.50/ft[d]	20/ft
Raft River and Boise					
Distribution piping	Transite carbon steel with external coating	8.60/ft[a,e] 25/ft[a]	PC or PIC PC or PIC	5/ft[a] 5/ft[a]	3.60/ft 20/ft

*The English units used in the body and footnotes of this table have the following metric equivalents:
1 in. = 25.4 mm;
1 ft = 0.3048 m;
1 lb = 0.4536 kg; and
1 psi = 6.8948 kPa.

[a] Assume a 12-in. pipe.
[b] Assume a 0.25-in. plate.
[c] Assume a 0.5-in. plate.
[d] Assume a 4-in. pipe.
[e] Pressure rating, 200 psi.

References

[1] Steinberg, M., "Concrete-Polymer Materials Development, a Goal Oriented Program," BNL 50313, Brookhaven National Laboratory, Upton, N.Y., 6 Oct. 1971.
[2] Kukacka, L. E. et al, "Concrete Polymer Materials, Fifth Topical Report," BNL 50390, Brookhaven National Laboratory, Upton, N.Y., Dec. 1973.
[3] Kukacka, L. E. et al, "Alternate Materials of Construction for Geothermal Applications," Progress Report No. 15, BNL 50834, Brookhaven National Laboratory, Upton, N.Y., Oct. 1977–March 1978.
[4] Kukacka, L. E. et al, "Concrete-Polymer Materials for Geothermal Applications," Progress Report No. 8, BNL 21244, Brookhaven National Laboratory, Upton, N.Y., Jan.–March, 1976.
[5] Kukacka, L. E., "Polymer-Concrete Composites for Energy Related Systems," Progress Report No. 4, BNL 19970, Brookhaven National Laboratory, Upton, N.Y., Jan.–March 1975.
[6] Sugama, T. and Kukacka, L. E., *Cement and Concrete Research*, Vol. 9, 1979, pp. 69–76.
[7] Zeldin, A., Kukacka, L. E., Fontana, J. J., Carciello, N. R., and Reams, W., *Journal of Applied Polymer Science*, Vol. 24, 1979, pp. 455–464.

A. N. Zeldin,[1] *L. E. Kukacka,*[2] *J. J. Fontana,*[3] *and*
N. R. Carciello[4]

Organosiloxane Polymer Concrete for Geothermal Environments

REFERENCE: Zeldin, A. N., Kukacka, L. E., Fontana, J. J., and Carciello, N. R., "**Organosiloxane Polymer Concrete for Geothermal Environments,**" *Geothermal Scaling and Corrosion, ASTM STP 717,* L. A. Casper and T. R. Pinchback, Eds., American Society for Testing and Materials, 1980, pp. 194–206.

ABSTRACT: The feasibility of using the products of free-radical copolymerization of modified organosiloxane in the formation of a thermally stable and chemically resistant polymer concrete for use in geothermal environments has been demonstrated. Specimens have been produced by using mixtures of organosiloxane containing pendant vinyl groups and styrene or different silicone fluids as a comonomer in conjunction with a free-radical initiator and several aggregate materials. The use of these monomers in conjunction with materials such as silicon dioxide (SiO_2) and portland cement to form polymer concrete results in composites with high compressive strength (80 to 100 MPa) and thermal and hydrolytic stability. The results from studies to determine the effect of variables, such as sand particle size, type of cement, and sand/cement ratio, are discussed.

KEY WORDS: geothermal, scaling, corrosion, concrete polymer materials, geothermal fluids, polymer-impregnated concrete, polymer concrete

For the last ten years, researchers from many countries have devoted a great deal of attention to the problem of developing high-strength, chemical-resistant cements that can be used for high-temperature applications. One type of material, which has been investigated at Brookhaven National Laboratory (BNL) since 1971 [1],[5] is formed by combining organic and

[1] Associate chemical engineer, Materials Systems Group, Process Sciences Division, Department of Energy and Environment, Brookhaven National Laboratory, Upton, N.Y. 11973.
[2] Leader of Materials Systems Group, Process Sciences Division, Department of Energy and Environment, Brookhaven National Laboratory, Upton, N.Y. 11973.
[3] Materials research scientist, Materials Systems Group, Process Sciences Division, Department of Energy and Environment, Brookhaven National Laboratory, Upton, N.Y. 11973.
[4] Chemistry associate, Materials Systems Group, Process Sciences Division, Department of Energy and Environment, Brookhaven National Laboratory, Upton, N.Y. 11973.
[5] The italic numbers in brackets refer to the list of references appended to this paper.

inorganic materials to produce a series of composites called concrete polymer materials. As a result of this work, two of the materials, polymer-impregnated concrete and polymer concrete, are beginning to be utilized throughout the world in applications where portland cement cannot be used or where severe maintenance problems occur. Extensive work on these composites is being conducted in Germany, Japan, and the USSR [2].

Polymer-impregnated concrete (PIC) consists of a precast portland cement concrete impregnated with a monomer system that is subsequently polymerized *in situ*. The polymer tends to fill the porous void volume of the concrete, which results in significant improvement in strength and durability. For a concrete mix that produces specimens with a compressive strength of 35.0 MPa, compressive strength >130.0 MPa has been measured after impregnation with the monomer. Similarly, large improvements in other structural and durability properties have also been obtained.

Polymer concrete (PC) consists of an aggregate mixed with a monomer or resin that is subsequently polymerized in place. The techniques used for mixing and placement are similar to those used for portland cement concrete. After curing, a high-strength (>75 MPa), durable material is produced.

The filler materials used in the production of PC include sand, limestone, slag, and expanded clay. Resins such as polyester, polyurethane, polymethyl methacrylate, polystyrene, and carbamide find application as the bonding agent.

Results from laboratory and field tests indicate that one of the areas in which these composites can be used is the area of geothermal processes [3]. The feasibility of these applications was demonstrated in 1972 when tests of a PC liner on a vertical tube evaporator in Freeport, Tex., indicated that the composites had long-term stability in seawater at 177°C and in acid solutions [4]. On the basis of these results, a research program to develop the composites for use in geothermal systems was started in April 1974. To date, several high-temperature PC systems have been formulated, and laboratory and field tests have been performed in brine, flashing brine, and steam at temperatures up to 250°C [3,5]. In all of the systems tested, organic resins were used as the binding agent.

It is well known that organic networks consist of carbon-carbon bonds, which are cleaved at temperatures near 250°C. As a result, the use of organic binders in PC is limited to temperatures below 250°C. Above this temperature, PC specimens become soft, swell, and crack because of the evolution of gas.

This paper deals with new semiorganic polymer compositions that have a network consisting of -Si-O-Si bonds. This bond has a dissociation energy at 25°C of 190.9 ± 2 kcal/mol, in comparison with the 145 ± 5 kcal/mol for the C-C bond. Also, the molecular chains of siloxanes are more rigid than those of organic polymers (the bond lengths of -Si-O- and -C-C are

1.504 and 1.541 Å, respectively). Thus, it is believed that siloxanes in composition with various fillers, in contrast to the unfilled polymers, should be stable at temperatures up to 350°C [7].

Experimental

Materials

The following materials were used in this study: dimethylpolysiloxane modified with pendant vinyl groups (Y-9208), supplied by the Union Carbide Corp.; silicone fluids RZ-4035 and V-47, supplied by the Rhodia Corp.; and styrene from the Dow Chemical Co. These materials were used as bonding materials, and sand and cement were used as fillers. The curing agent for the polymerization was di-*tert*-butyl peroxide (DTBP).

Specimen Preparation

Polymer concrete specimens were made by blending the comonomers at 25 to 30 weight percent and a free-radical initiator at 0.5 to 1.0 weight percent, and the formulation was mixed with the appropriate aggregate mixture. The slurry was then evacuated to remove entrapped air. The monomer concrete slurry was vibrated into 25-mm-diameter glass cylinders and cured in these cylinders. When the PC was removed from the glass tubes, it was cut into cylinders that measured 25 mm in diameter by 50 mm in length. These cylinders were used in measurements of properties.

Test Methods

The water absorption and compressive strength tests for the PC specimens were performed in accordance with ASTM standards [6]. The absorption indicates the percentage of liquid that is drawn into and tends to fill permeable pores in a porous solid body and is calculated as follows

$$\frac{W_2 - W_1}{W_1} \text{ (100 percent)}$$

where

W_1 = weight of the dried specimen before the test, and
W_2 = weight of the specimen after the test.

Compressive strength tests were performed on PC samples before and after exposure to 25 percent brine solutions at elevated temperatures.

Durability tests in autoclaves containing a simulated geothermal brine were also performed. In these tests the specimens were first weighed and dimensionally measured. They were then installed in high-pressure autoclaves containing 25 percent brine solutions at room temperature. The autoclaves were brought up to temperature, and the pressure was recorded. After the desired testing time had been reached, the vessel was opened, and the specimens were washed in water (H_2O) and dried before being tested.

The analytical equipment used to evaluate the PC included an AMR Model 1000A scanning electron microscope equipped with LaB_6 and a Perkin-Elmer infrared (IR) spectrophotometer, Model 297. The potassium bromide pellet procedure was used for the IR analysis.

Results and Discussion

An earlier paper [7] discussed the possibility of using organosiloxanes with organics or different siloxane fluids as a means of producing PC that was mechanically and chemically stable at high temperature. This work indicated that, in contrast to PC that contained organic polymers, which showed a marked decrease in strength with temperature, the organosiloxane-styrene PC decreased in strength at a much slower rate, and at 350°C the compressive strength was ~25 MPa (see Fig. 1).

The goal of the current study was to optimize the monomer and aggregate compositions and to determine the properties of the PC as a function of sand particle size, type of cement, monomer composition, sand/cement ratio, and concentration of the coupling agent. Partial results from these studies are reported here.

Sand Particle Size

As shown earlier [8], the particle size of the sand has an effect on the properties of PC, and this effect is dependent upon the viscosity of the monomer composition. The best results for the organic binders with viscosities close to the viscosity of water were obtained when a mixture of sand sizes was used [8]. In this case, the low-viscosity monomer could easily be mixed with the filler to wet the particles and to fill the voids in the aggregate. When the viscosity of the monomer mixture is increased, the percentage of filler that the monomer wets decreases. The result, therefore, is nonuniform polymer loading, which decreases the durability and strength of the PC. Table 1 illustrates the effect of the sand particle size on the compressive strength of PC composites. Decreasing the particle size of the sand for a 75 weight percent organosiloxane/25 weight percent styrene system increased the strength of the PC specimens from 4.7 to 41.6 MPa. It is also apparent from the data in Table 1 that the addition of portland cement results in improved strength. When 10 weight percent Type III

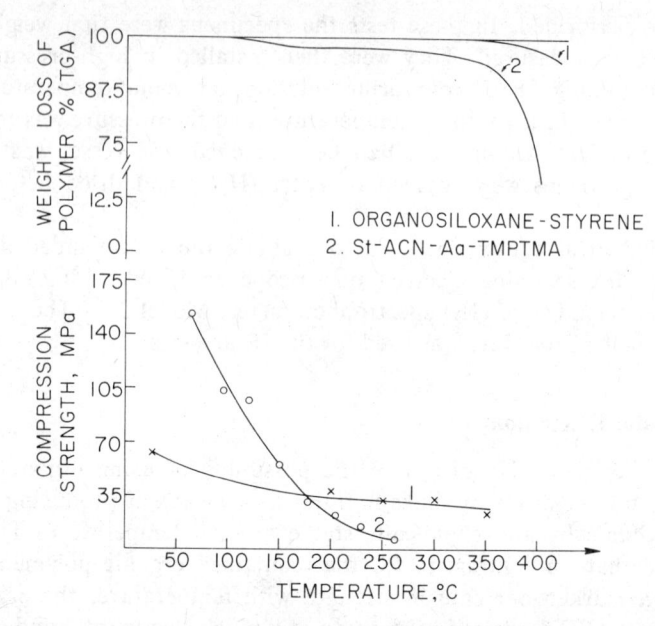

FIG. 1—*Polymer weight loss and PC compressive strength at various temperatures. St = styrene; ACN = acrylonitrile; Aa = acrylamide; TMPTMA = trimethylolpropane trimethacrylate.*

TABLE 1—*Compressive strength of organosiloxane as a function of sand particle size.*

Monomer Composition, Weight Percent[a]		Aggregate Composition		Compressive Strength, MPa
OS[b]	St[c]	Weight Percent	Sieve Opening, μm	
75	25	100	<30	41.6
75	25	100	150	12.5
75	25	100	600	4.7
75	25	90 10 Type III cement	<30	85.2
75	25	90 10 Type III cement	150	31.6
75	25	90 10 Type III cement	600	21.8
75	25	36[d] 17 17 30 Type III cement	1180 600 150	40.3

[a] Monomer loading, 35 weight percent.
[b] OS = organosiloxane.
[c] St = styrene.
[d] = aggregate composition, used for organic systems.

portland cement was added to the filler, the compressive strength of the material increased to 85.2 MPa.

Type of Cement

As has been indicated, the addition of cement significantly improves the strength and durability of siloxane-PC (SPC) composites. A study to determine the influence of various types of cement on the properties of PC is in progress. Four different, commercially available types of cement, which were included as a partial constituent of the aggregate, were tested: Types I, II, III, and V. The compound compositions [9] of these materials are given in Table 2.

The properties of PC as a function of cement type are summarized in Table 3 for a 75 weight percent Y-9208/25 weight percent styrene monomer system and in Table 4 for a 97 weight percent RZ/3 weight percent V-47 system. Polymer concrete specimens containing Types I and III cement in conjunction with the Y-9208/styrene monomer system had high strengths (Fig. 2). Type II cement used with the RZ/V-47 system also yielded high strength (Fig. 3). The sand/cement ratios were 90/10 and 95/5, respectively. The amount of water absorbed was lowest for specimens containing Types III and V cement (Figs. 2 and 3). Stress-strain curves (Figs. 4 and 5) indicate that PC specimens containing Types I and II cement are more brittle than PC containing the other cements. This brittleness can be reduced slightly by using Type III cement. Greater reductions can be obtained with Type V cement. The reasons for these variations may be the differences in the chemical composition of the cements. The feasibility of increasing the elasticity of PC could be useful in developing high-temperature sealing materials for the geothermal industry.

Sand/Cement Ratio

Studies are also in progress to determine the optimum ratio of sand to cement for the PC composites. The results from compressive strength tests

TABLE 2—*Compound composition of portland cements.*

Type of Cement	C_3S^a	C_2S^a	C_3A^a	C_4AF^a	$CaSO_4$	Free CaO	MgO	Ignition Loss
I	49	25	12	8	2.9	0.8	2.4	1.2
II	46	29	6	12	2.8	0.6	3.0	1.0
III	56	15	12	8	3.9	1.3	2.6	1.9
V	43	36	4	12	2.7	0.4	1.6	1.0

[a] In the abbreviated formula C = CaO; S = SiO_2; A = Al_2O_3; F = Fe_2O_3.

TABLE 3—*Properties of PC as a function of cement type.*

Aggregate Composition, Weight Percent		Cement Type	Boiling H$_2$O Absorption, %	Compressive Strength, MPa	Modulus of Elasticity, MPa
Sand[a]	Cement				
90	10	I	0.075	111.8	6500
		II	0.072	46.6	4301
		III	0.03	119.3	5175
		V	0.02	100.7	4825
80	20	I	0.12	66.2	...
		II	0.05	63.8	...
		III	0.06	42.5	...
		V	0.07	95.1	...
70	30	I	0.09	60.0	...
		II	0.08	105.3	...
		III	0.12	73.8	...
		V	0.05	104.2	...

[a] Characteristics of sand:
Particle size: sieve opening <30 μm.
Monomer composition: 75 weight percent Y-9208/25 weight percent styrene.
Curing agent: 1.0 weight percent di-*tert*-butyl peroxide.
Curing condition: 110°C, 18 h; 165°C, 4 h.
Monomer concentration: 35 weight percent.

TABLE 4—*Properties of PC as a function of cement type.*

Aggregate Composition, Weight Percent		Cement Type	Boiling H$_2$ Absorption, Percent	Compressive Strength, MPa	Modulus of Elasticity, MPa
Sand[a]	Cement				
95	5	I	0.06	81.0	...
		II	0.09	111.2	...
		III	0.1	70.5	...
		V	0.09	49.1	...
90	10	I	0.07	106.9	6826
		II	0.08	101.6	...
		III	0.05	52.9	6791
		V	0.11	74.39	7526
85	15	I	0.11	86.1	...
		II	0.08	102.9	...
		III	0.08	72.2	...
		V	0.13	70.85	...
80	20	I	0.22	47.5	...
		II	0.12	71.6	...
		III	0.19	72.7	...
		V	0.14	68.2	...

[a] Characteristics of sand:
Particle size: sieve opening <30 μm.
Monomer composition: 97 weight percent RZ/3.0 weight percent V-47.
Curing agent: 1.0 weight percent di-*tert*-butyl peroxide.
Curing condition: 110°C, 18 h 135°C, 3 h; 170°C, 2 h.
Monomer concentration: 35 weight percent.

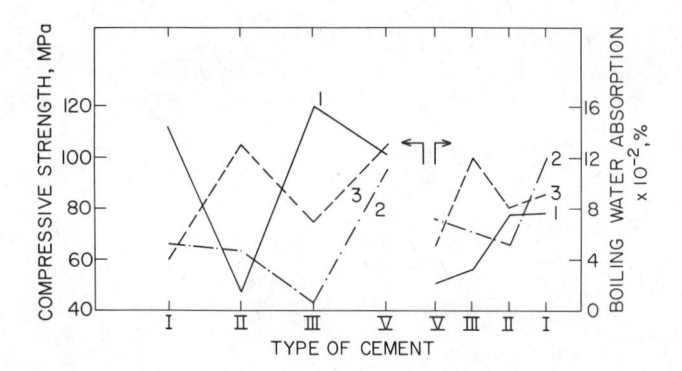

FIG. 2—*Compressive strength and boiling water absorption as a function of cement type. Monomer composition 75 weight percent Y-9208/25 weight percent styrene. Sand/cement ratio: (1) 90/10, (2) 80/20, and (3) 70/30.*

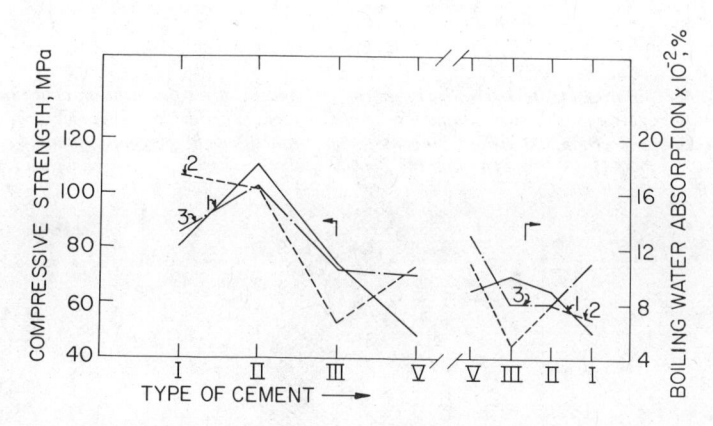

FIG. 3—*Compressive strength and boiling water absorption as a function of cement type. Monomer composition: 97 weight percent RZ/3 weight percent V-47. Sand/cement ratio: (1) 95/5; (2) 90/10; and (3) 85/15.*

made before and after the exposure of specimens in autoclaves to brine at temperatures of 275 and 300°C are summarized in Table 5 and Fig. 6. The data indicate that PC specimens containing cement up to a concentration of 15 percent are stable after exposure to brine at 275°C. Only the composites containing sand and cement in ratios of 95/5 and 90/10 showed stability at 300°C.

Infrared spectroscopy analyses (Fig. 7) and scanning electron microscopy studies (Fig. 8) of the structure of PC specimens before and after exposure to brine at 300°C did not indicate changes in structure and composition. As indicated in Fig. 8, some fragments of the resin adhering to the filler

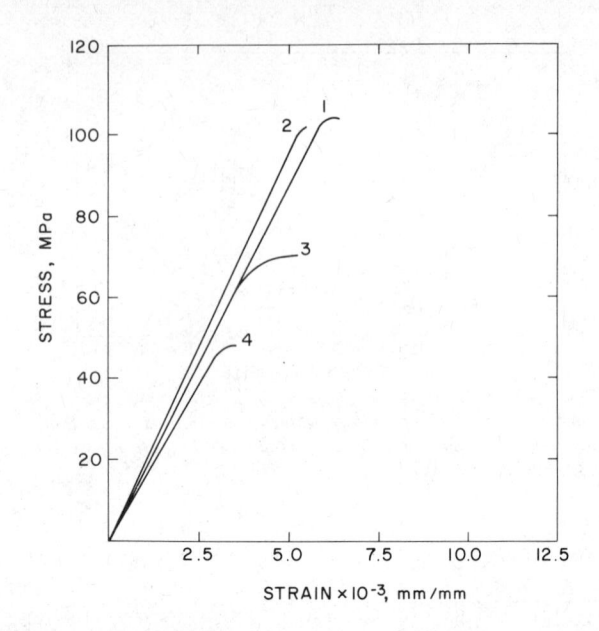

FIG. 4—*Compressive stress-strain curves for siloxane PC after preparation as a function of cement type. Monomer composition: 75 weight percent Y-9208/25 weight percent styrene. Aggregate composition: 90 weight percent flour sand/10 weight percent cement. (1) Type I cement; (2) Type II; (3) Type III; and (4) Type V.*

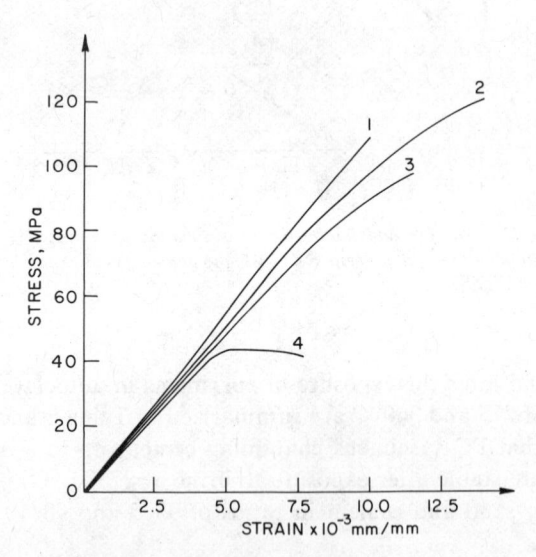

FIG. 5—*Compressive stress-strain curves for PC after preparation as a function of cement type. Monomer composition: 97 weight percent RZ/3 weight percent V-47. Aggregate composition: 90 weight percent flour sand/10 weight percent cement. (1) Type I cement; (2) Type II; (3) Type III; and (4) Type V.*

TABLE 5—*Sand/cement ratio for System RZ/V-47 (ratio 97/3).*

Aggregate Ratio, Weight Percent		Boiling Water Absorption, Weight Percent	Compressive Strength, MPa, After			Modulus of Elasticity, MPa After Exposure to Hot Brine for 30 Days at 300°C
Sand[a]	Cement[b]		Boiling Water	30 Days in Autoclave at 275°C	30 Days in Autoclave at 300°C	
100	0	0.02	49.9	48.3	...	...
95	5	0.08	74.2	55.9	79.6	8845
90	10	0.07	90.2	69.7	57.1	7963
85	15	0.04	36.2	40.6	...	...
80	20	0.54	15.2	weak	...	...
70	30	0.69	34.2	weak	...	...
95[c]	5	0.09	89.1	48.5	60.9	9100
90[c]	10	0.12	46.4	107.2	62.8	5460

[a] Particle size: sieve opening <30 µm.
[b] Portland cement Type III.
[c] These samples were made by using wetting agent.
Monomer concentration: 35 weight percent.
Initiator: ½ weight percent DTBP; ½ weight percent Silane A-174.
Curing condition: 125°C, 16 h; 150°C, 3 h; 180°C, 3 h.

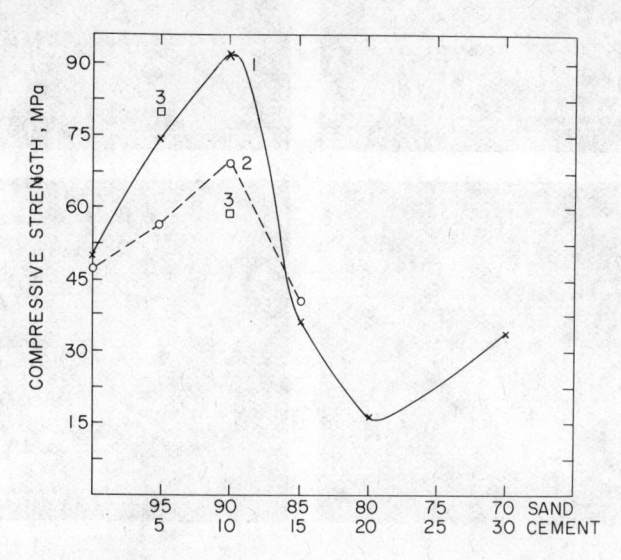

FIG. 6—*Compressive strength as a function of sand/cement ratio. Monomer composition: 97 weight percent RZ/3 weight percent V-47. Aggregate composition: sand-sieve opening <30 µm. Cement: Type III. Compressive strength: (1) after boiling water; (2) after exposure to brine at 275°C; and (3) after exposure to brine at 300°C.*

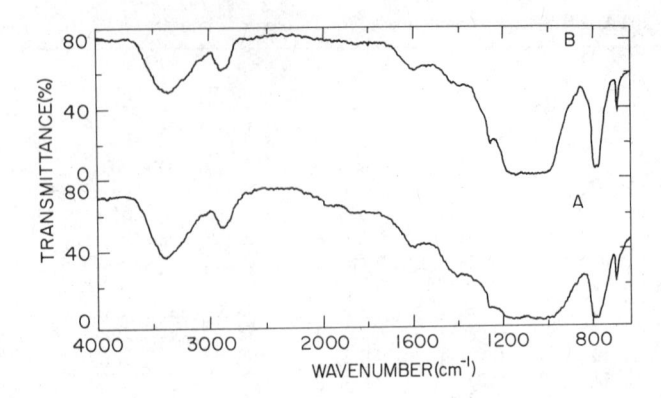

FIG. 7—*Infrared absorption spectra of the PC samples before* (A) *and after* (B) *exposure to brine at 300°C.*

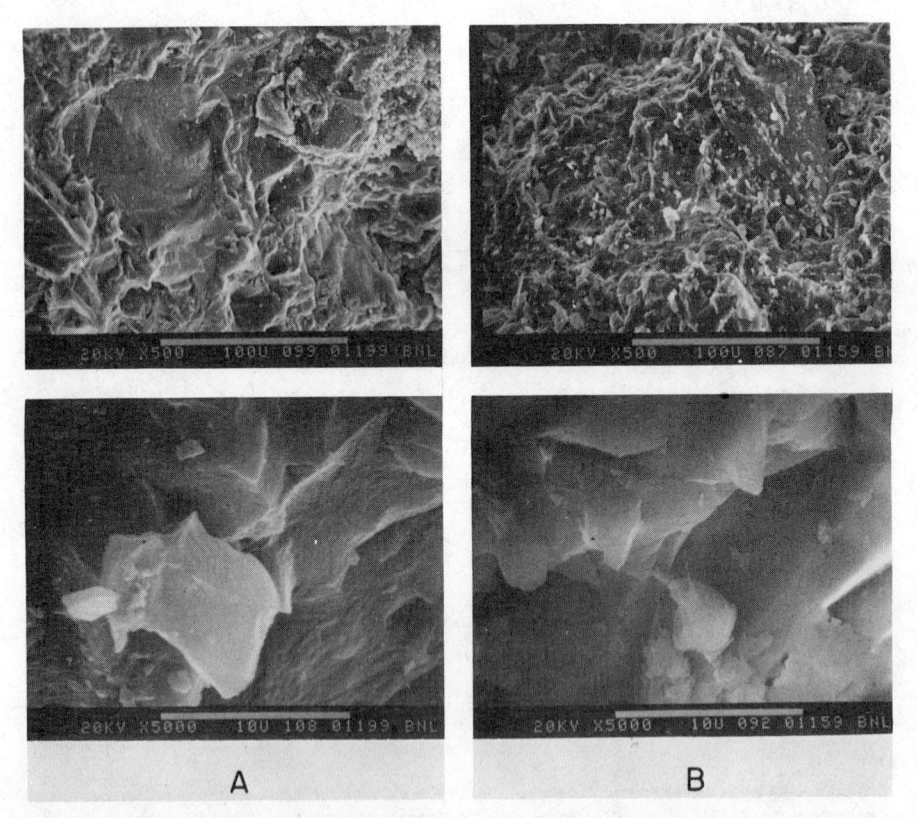

FIG. 8—*Scanning electron microscope pictures of the fractured surface of PC specimens before* (A) *and after* (B) *exposure to brine at 300°C.*

surface have been observed. These observations support the presence of adhesion between the polymer and filler. The boiling water absorption, which is representative of the PC density, has a tendency to increase with increasing amounts of cement. Figure 6 indicates that maximum strength is obtained when the cement concentration used is in the range of 10 to 15 weight percent. Further increases in the cement concentration lead to strength reductions and increases in elasticity (Table 5). The use of a wetting agent (Triton X-100 from the Baker Chemical Co.) does not change this tendency (see Table 5 and Fig. 9) but improves the particle-particle packing of the various mixtures of monomer and aggregate.

Summary and Conclusion

A study has been performed to determine the suitability of using copolymers obtained by the free-radical polymerization of mixtures of organosiloxane with different silicone fluids or with styrene to produce thermostable polymer concrete composites for use in geothermal environments. The properties of the PC were measured as a function of sand particle size, cement type, and sand/cement ratio for 75 weight percent Y-9208/25 weight percent styrene and 97 weight percent RZ/3 weight percent V-47 monomer systems. The optimum properties result when sand flour with a sieve opening <30 μm in diameter is used in conjunction with Type II or III cement in a ratio of 90/10 or 95/5.

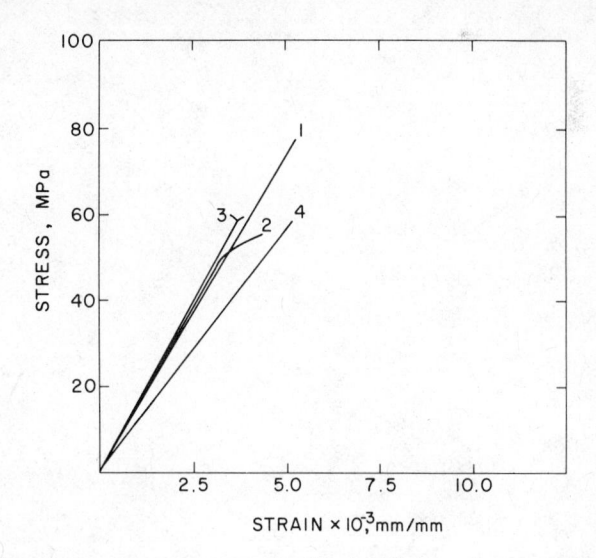

FIG. 9—*Compressive stress-strain curve for PC after 30 days of exposure to brine at 300°C. Monomer composition: 97 weight percent RZ/3 weight percent V-47. (1) Sand/cement ratio 95/5; (2) sand/cement ratio 90/10; (3) the same as (1) plus a wetting agent; and (4) the same as (2) plus a wetting agent.*

Acknowledgment

This work was performed under the auspices of the U.S. Department of Energy under Contract No. EY-76-C-02-0016.

References

[1] Steinberg, M., Colombo, P., Kukacka, L. E., and Manowitz, B., U.S. Patent No. 3,567,496, 1971.
[2] *Proceedings,* Second International Congress on Polymers in Concrete, Austin, Texas, Oct. 1978.
[3] Kukacka, L. E., "Polymer Concrete Materials for Use in Geothermal Energy Processes," Second International Congress on Polymers in Concrete, Austin, Texas, Oct. 1978.
[4] Kukacka, L. E., Fontana, J., Kukacka, L., and Carciello, N., "Alternate Materials of Construction for Geothermal Applications," Progress Report No. 14, BNL 50751, Brookhaven National Laboratory, Upton, N.Y., Sept. 1977.
[5] Zeldin, A., Kukacka, L. E., and Carciello, N., *Journal of Applied Polymer Science,* Vol. 23, 1979, pp. 3179–3191.
[6] *1978 Annual Book of ASTM Standards, Part 14,* American Society for Testing and Materials, Philadelphia, 1978.
[7] Zeldin, A., Fontana, J., Kukacka, L., and Carciello, N., *International Journal of Cement Composites,* Vol. 1, 1979, pp. 11–17.
[8] Steinberg, M. "Concrete-Polymer Materials," Progress Report No. 3, BNL 50275 (T-602), Brookhaven National Laboratory, Upton, N.Y., January 1971.
[9] *Concrete Manual,* U.S. Bureau of Reclamation, Government Printing Office, Washington, D.C., 1975.

S. L. Phillips,[1] *A. K. Mathur,*[2] *and Warren Garrison*[3]

Treatment Methods for Geothermal Brines*

REFERENCE: Phillips, S. L., Mathur, A. K., and Garrison, Warren, "**Treatment Methods for Geothermal Brines,**" *Geothermal Scaling and Corrosion, ASTM STP 717,* L. A. Casper and T. R. Pinchback, Eds., American Society for Testing and Materials, 1980, pp. 207-224.

ABSTRACT: A survey is made of commercially available methods currently in use, as well as those which might be used, to prevent scaling and corrosion in geothermal brines. More emphasis is placed on scaling than on corrosion. Treatments are classified as inhibitors, alterants, and coagulants and are applied to control scaling and corrosion in fresh and waste geothermal brines. Recommendations for research in brine treatment are described.

KEY WORDS: geothermal energy, brines, brine treatment, scaling, scale prevention, corrosion prevention, inhibitors, coagulants

During the past 10 years, much progress has been made in the United States in developing geothermal energy and constructing power plants. The current electrical power produced is 608 MWe at the Geysers in Calif., which obtains steam to drive turbines from steam wells. However, the major new sources of geothermal energy in the next decade are expected to be hot brine systems located in western United States. A problem in utilizing these hot-water resources is the dissolved gases and minerals present, which cause scaling and corrosion of wells, piping, heat exchangers, and other components of the power plant [1].[4]

Methods for treating brines to control scaling and corrosion in production and injection wells, and within the power plant, are needed: (1) to permit

---
*Any conclusions or opinions expressed in this report represent solely those of the authors and not necessarily those of the Regents of the University of California, the Lawrence Berkeley Laboratory, or the U.S. Department of Energy.
[1]Chemist, Lawrence Berkeley Laboratory, University of California, Berkeley, Calif. 94720.
[2]Petroleum engineer, Texaco, Inc., Los Angeles, Calif. 90010.
[3]Chemist, Lawrence Berkeley Laboratory, University of California, Berkeley, Calif. 94720.
[4]The italic numbers in brackets refer to the list of references appended to this paper.

utilization of sites with high scaling potential, such as the Salton Sea geothermal fluids, which contain as much as 250 000 ppm total dissolved solids, and (2) to lower the cost of maintenance of power plants by reducing the rate of corrosion and formation of scale at each site.

The control of scale formation and corrosion has a long history of careful empirical testing, research, and maintenance. Scale is typically removed by scraping and reaming; often acid is added prior to scraping, and the products are removed by washing with water. These are effective methods for removing scales once they have formed [2].

Currently used methods of controlling scaling in geothermal plants are centered on mechanical means, such as using wire brushes to remove scale from pipes and acidizing or reaming wells. Scale removal is considered a maintenance problem and scheduled as needed.

Since about 1950, research has been in progress in the oil-field and boiler water areas to develop means of controlling scale and corrosion by adding chemical substances. Treatment methods are devised for both the fresh fluid (for example, water, petroleum), prior to its use in power generation or refining, and the spent brines, prior to their injection into the ground or disposal by other methods. The treatment methodology of geothermal brines parallels that for oil-field and other brines, and those portions that appear relevant are included here.

The increasing availability of basic data for the solubility and kinetics of corrosion and scaling parameters, as well as the development of computer methods, has made it possible to examine increasingly complex scaling problems. The computer calculations are based on solubility and equilibrium data for mineral solutions and other substances, as well as on flow rates and temperatures.

Treatment methods to control geothermal and other scaling can be classified in a number of ways. It is convenient to separate them into the following categories: (1) inhibitors, (2) alterants, and (3) coagulants. Inhibitors are substances added to brines, usually at the parts-per-million level, to retard the growth of scale; they are generally added to the fresh brine. Alterants are added to the brine, with a resulting change in the chemistry; examples are the addition of hydrochloric acid to lower the pH, and the removal of metals by ion exchange. The third class covers coagulants and flocculants, which are added to remove precipitates or suspended substances; they are generally added to wastewaters as a pretreatment method prior to disposal, for example, by injection. Table 1 lists these three classes of methods commonly used to treat geothermal and other waters.

It is not possible to cover both scaling and corrosion in detail in this report. Although emphasis here is on geothermal scaling, research in other areas such as oil-field scaling, which might be applicable to the control of geothermal scale formation, is also included. In this report, we cover scale formation and treatment for both the fresh brine from production wells, which is

TABLE 1—*Typical treatment methods to prevent scaling of fresh and spent geothermal and other brines.*

Treatment Method	Prevents or Controls
Inhibitors	
Lime slurry	calcite
Phosphonate + polymer	silica scale, mixed scales
Ethylene oxide polymer	silica, corrosion
Hydroxethylcellulose	silica deposition, corrosion
Low molecular weight carboxylic acid	silica, corrosion
Amine	silica, corrosion
Sludge	silica, corrosion
Phosphoric acid	calcite
Polymeric carboxylic acid	calcite
Seeding with scale	cacite
EDTA	calcite
Polyacrylate	scales
Dispersant (highly carboxylated polymers)	calcite
Phosphate + sand	calcite, $BaSO_4$, $CaSO_4$
Solutions of amines, amides, carboxylic acids	scales, inhibits to 204.4°C (400°F)
Molybdate	scales
Alterants	
Hydrochloric acid	silica, calcite
Hydrogen peroxide, nitric acid	H_2S
Fresh water diluent	silica
Heavy diesel oil	silica, FeS, borate
CO_2 pressure	calcite
Temperature	silica, calcite
Air, oxygen	H_2S, Mn^{++}, Fe^{++}
Chlorine	H_2S, biogrowth
Ion-exchange resin	dissolved metals, borate
Coagulants and Flocculants	
Anionic polyelectrolyte	flocculant
Slaked lime + hypochlorite	silica, arsenic
Aluminum sulfate, ferrous sulfate	suspended solids, colloids
Cellulose xanthate	heavy metals
Other	
Metallic core piping	scale, corrosion
Ultraviolet light	biogrowth

flashed to steam and used to drive a turbine, and the spent fluid, which is treated prior to disposal, for example, by way of injection wells. More emphasis is placed on the treatment of fresh fluids, where the current state of the art is not as well developed as it is for spent brines.

Geothermal Scale

Scale formation is common in geothermal hot water systems and arises from deposition and adhesion of soluble or suspended constituents such as silica and calcite onto pipes, walls, and other components of power plants. The need to control scaling stems from three major difficulties: (1) the plugging and clogging of wells and pipes transporting the brine, (2) the decrease in the efficiency of turbines and heat exchangers, and (3) the freezing of valves, which renders instruments inoperative. The second problem covered here is corrosion of metallic components in contact with the hot brine. Geothermal brines contain dissolved gases and salts, such as carbon dioxide (CO_2), hydrogen sulfide (H_2S), and sodium chloride (NaCl), that are generally more corrosive to materials of construction than other environments common to power production.

Table 2 shows that the scale constituents commonly found in geothermal systems are silicate, carbonate, and sulfide. Depending on the power plant location, the source of the scale is fresh brine, flashed steam from the brine, or spent fluid condensate. The composition of typical brines is shown in Table 3.

Brine Treatment Systems

The developers of systems for treating geothermal brines are faced with a large selection of possible treatment additives, corrosion-resistant materials, and disposal options. The selection of an optimum treatment method is further complicated by the combined needs of preventing scale while not affecting the temperature and flow rate of the production fluid. The waste brines must be treated both for compatibility with the injection well and receiving formation and for preventing undesirable pollutants (for example, boron, arsenic) from mixing with surface waters. Besides these, large quantities of brine and solids such as sludge must be treated and disposed of.

The developer needs to consider that in treatment systems the following basic functions must be provided: chemical analysis, holding tanks, coagulants and filters, pumps, and associated piping. Each function is important, and the entire treatment system must be considered in order to realize the full operation capabilities of the system. The discussion which follows applies mainly to the step in which material, for example, an inhibitor or coagulant, is added to the brine.

Besides the addition step, one or more of the following components will be needed by a system: sampling methods for fresh and spent brine, samples of steam and condensate, pipes to carry the inhibitor or alterant; mixing and storage tanks, coagulation and sedimentation facilities, sludge disposal sites, pumps for injection wells and for the addition of additives, instrumentation to measure pH and other important parameters, chemical analysis of the waste brine and fresh brine, regulatory requirements, data on the injection (receiving) formation, and fluid flow rates.

Both the discussion that follows and specific examples given will center on the more important geothermal scale components: silica, calcite, and sulfide. The treatment methods covered include those currently used in geothermal plants as well as those commonly used in other industries [3,4,5,6].

TABLE 2—*Scale compositions commonly found in selected geothermal sites.*

Site	CaCO$_3$	CaO	SiO$_2$	S	SO$_4$	Fe	Other
Matsukawa, Japan							Al
Well No. 1 (pH 5.0)	...	0.30	17.75	3.20	40.84	12.20	0.83
Well No. 2 (pH 7.5)	...	0.59	90.45	...	2.25	0.35	0.84
Power plant generator	...	1.5 to 1.7	59 to 61	0.05 to 53	10.2 to 13.8	3.8 to 46.9	...
Control valve	...	1.2 to 1.60	44.4 to 69.28	...	17.56 to 32.82	1.00 to 2.01	...
Ejector	...	0.49	40.6	2.1	15.5	27.4	...
Cerro Prieto, Mexico							FeS
Well casing, M-5	1.5	...	15.1	...	...	...	83.4
Well casing, M-7	93.0	...	1.8	...	...	...	1.2
Production pipe, M-9	75.56	...	12.51	...	...	...	9.46
Otake, Japan							Fe$_2$O$_3$
Disposal pipe entrance	...	...	78.45	...	...	...	3.22
3863 m from entrance	...	...	93.25	...	...	...	0.15
East Mesa, Calif.							
Well 38-30	major	...	...	...	...	...	...
Vertical tube evaporator	major	...	...	...	...	...	Ca
Salton Sea/Niland, Calif.							
Teflon coupon	...	...	83	0.3	...	...	0.45
Heber, Calif.							Si
Heat exchanger, steel	...	...	...	19 to 41	...	18 to 62	1 to 21
Heat exchanger, titanium	...	...	...	15 to 45	...	0.5 to 4	1 to 11
Lardarello, Italy							FeS
Turbine	present	...	...	...	present	...	present
Hveragerdi and Namafjall, Iceland							Fe$_2$O$_3$
Heat exchanger	...	...	61 to 73	...	...	...	0.3 to 4

TABLE 3—*Typical concentrations of key chemical species in fluid from seven geothermal sites.*

KGRA	Temperature, deg C, and location	pH	Cl⁻	Total CO_2	Total H_2S	Total NH_3	$SO_4^=$	Fluid Description	SiO_2	TDS
Salton, Sea, Calif.	250 (borehole)	5.2	115 000	1000	10 to 30	300	20	unflashed wellhead fluid	500	120 000 to 250 000
East Mesa, Calif.	180 to 200 (borehole)	5.7	11 000	800	3	41	20	unflashed wellhead fluid	330	3 000
Heber, Calif.	180 to 200 (borehole)	7.1	9 000	180	~2	13	152	unflashed wellhead fluid	...	14 000
Mono-Long Valley, Calif.	175 (borehole)	6.5	227	180	14	0.1	96	unflashed wellhead fluid	...	...
Baca (Valles Caldera) N. Mex.	171 (wellhead at 7.6 bar)	6.8	3 770	128	6	...	59	flashed fluid	...	...
Beowawe, Nev.	132 (wellhead)	9.3	50	209	6	3	89	flashed fluid	329	1 200
Raft River, Idaho	146 (borehole)	7.2	780	60	0.1	2	61	unflashed fluid	...	1 204 to 3 360

Properties of the Geothermal Resource
Concentration of Key Species in the Fluid, ppm

Scale Inhibitors and Seeding

Prevention of scale formation by addition of inhibitors to fresh brines is largely done on a trial-and-error basis. Both proprietary and other chemicals are screened by treating the brine and noting the effect on the thickness of the resulting scale or any change in the scale composition [7,8]. The methods commonly used are seeding and adding scale inhibitors.

Seeding

The purpose of seeding is to initiate precipitation of the scale-forming substance at selected places under controlled conditions, thereby removing these components before they can deposit as scale in undesired locations. In this treatment method, a finely divided solid is introduced into the brine to help crystallize the soluble material. The method has been used for seawater [9] and geothermal brines [10,11], and in basic studies on seeding with calcium carbonate ($CaCO_3$) [12,13].

At the East Mesa area in the Imperial Valley, Calif., seeding was considered in fresh brine as a means of preventing scaling in process equipment, including a vertical tube evaporator. The method which involved the addition of lime and magnesium oxide to precipitate $CaCO_3$, magnesium hydroxide [$Mg(OH)_2$], and silica, was judged not attractive because the brine temperature would drop and large amounts of sludge would be generated—estimated at 270 000 to 540 000 kg (300 to 600 tons) per day for a 3.8 million-litre (1 000 000-gal) per day flow rate [11].

At the Niland geothermal area in Calif., a wet sludge suspension, composed of 20 percent by weight of sediment (mainly silica), from the geothermal loop experimental facility (GLEF) was metered into the fresh brine at 210°C, and the concentration of dissolved silica was measured. Before the addition of the sludge, the silica was 453 mg/kg; the sludge was metered in at 9000 litres/min (0.3 gal/min) and 4500 litres/min (0.15 gal/min), with the silica contents in the treated brine measured as 431 and 416 mg/kg, respectively [10].

Inhibitors

As used here, the term "inhibitor" refers to a substance which, when present at parts-per-million concentration levels, significantly retards the growth of scale or the formation of a precipitate. The inhibitors are tested and ranked in order of effectiveness and cost at high temperatures. This testing and ranking procedure must be done at each site because the brine properties and flow conditions vary for each geothermal site.

The inhibitor is generally thought to prevent deposits by preferentially forming a film on the growing scale crystal, thereby retarding further growth of the crystal [14]. For example, CaCO₃ deposition is prevented by phosphate inhibitors, which adsorb on the surface. Table 4 lists commercially available inhibitors[5] that have been used to control scaling in geothermal systems and selected inhibitors that are applied in other areas and that appear useful in geothermal energy utilization. In Ref 8 there is a listing of over 30 inhibitors tested for their effectiveness in inhibiting the precipitation of silica from hypersaline geothermal brines.

Inhibitors of the film type are generally polyphosphates (for example, sodium pyrophosphate), salts of acrylic and methacrylic acids, and aminoethyl and aminoethylene phosphonates [11].

Scale inhibitors are commonly classified as chelating agents, threshold inhibitors, and crystal distortion inhibitors [15]. Depending on the pH and the presence of anions, chelating agents such as ethylenediaminetetraacetic acid (EDTA) are effective for preventing scale due to calcium and iron by forming strongly bound soluble metal complexes. However, the stability of the metal-EDTA complex at geothermal temperatures and pressures, and in saline solutions, should be determined.

Threshold inhibitors are chemicals such as phosphonates added to the treatment water at concentrations just sufficient to retard scale formation. They are the most widely used inhibitors. Crystal distortion inhibitors limit the size of the scale crystals by producing small particles that remain suspended in the water. An example is the addition of Pb^{++} to distort the growth of $CaCO_3$ by forming lead carbonate ($PbCO_3$) [16].

Alterants

In the altering method of scale control, a brine property is modified by continuous or semicontinuous addition of a selected substance. Ideally, the brine property (for example, pH) is changed from its initial value to another value wherein the scale or corrosion is prevented. Alterants are gases, liquids, or ion-exchange solids (see Table 5).

In an experiment at Niland, Calif., hydrochloric acid (HCl) injection was found to be beneficial in controlling silica scale formation in brine from Magmamax Well No. 1. The pH of the unmodified brine was 5.5 to 5.8. The dissolved solids content of the brine prior to and after expansion through nozzles was 18 to 22 percent by weight. Nominal scaling, composed of copper sulfide, native silver, and iron-rich amorphous silica from the unmodified brine, resulted in closure of up to 10 percent of the cross-sectional areas of the nozzle throats. Scale formed on wear blades, ranging in thickness from 0.019 mm to 0.04 mm. However, when the brine was acidified to pH 1.5, 2.3,

---

[5] Reference to a company or product name does not imply approval or recommendation of the product by the University of California or the U.S. Department of Energy to the exclusion of others that may be suitable.

TABLE 4—*Typical commercially available scale inhibitors for geothermal and other waters.*

Manufacturer	Inhibitor	Comment
American Can Co.	Marasperse N-3	cellulose compound
American Hoechst	Tylose MHB	cellulose compound
Dow Chemical Co.	XFS-43075	
Union Carbide Corp.	Cellosize QP-09-L, Carbowax 350	hydroxyethylcellulose, ethylene oxide polymer
A. E. Staley	Starpol 100	substituted starch
Air Products	Surfynol 485	acetylenic glycol
E. I. DuPont	Zonyl FSN, FSB	fluorinated surfactants
Drew Chemical Corp.	Drewplex 502	tested at East Mesa
Calgon Corp.	Calgon CL-77W	tested at East Mesa
	Calgon SL-500	chelating type, tested at East Mesa
Monsanto Chemical Co.	Dequest 2060 (diethylenetri-amine-pentamethylenephosphoric acid)	tested at East Mesa
Hercules, Inc.	Natrosol 250LR (hydroxyethylcellulose)	tested at Salton Sea—Niland
Dearborn Chemical Corp.	Geomate 256 (phosphonate + polymer), Geomate 259	tested at Salton Sea—Niland
Calgon Corp.	CL-165 (polymer mixture)	tested at Salton Sea—Niland
Drew Chemical Corp.	Drewsperse 747 (phosphonate + polymer)	tested at Salton Sea—Niland
Betz Laboratories	Betz 419 (phosphonate + acrylic polymer), Polysperse Plus	tested at Salton Sea—Niland
Far-Best Corp.	Thermosol APS (polyalkylphosphonate)	tested at Salton Sea—Niland
C-E Natko	S-404 (organic polymer)	tested at Salton Sea—Niland
Southwest Specialty Chemicals	SC-210 (low molecular weight carboxylic acid)	tested at Salton Sea—Niland
Champion Chemicals	Cortron R-16 (filming amine)	tested at Salton Sea—Niland
National Starch & Chemical	Versa-TL3	used in boilers
Magma Corp. (Aquaness Div.)	Calnox Polyacrylates (214DN)	carbonate and sulfate scale control
Wright Chemical Corp.	Molybdate base	scale control
CIBA-GEIGY	Belgard EV (polymeric carboxylic acid)	seawater
Allied Colloids, Inc. (General Industries Div.)	ANTIPREX A (sodium polyacrylate)	developed for boilers and evaporators
Hercules, Inc.	SP-944, lime slurry additive	prevents $CaCO_3$
Colloid-A-Tron, Inc.	Colloid-A-Tron	piping which contains metallic core to prevent scaling and corrosion
Petrolite Corp. (Tetralite Div.)	Tolsperse 133	organic polyphosphonate
Nalco Chemical	NALCOOL	dispersive corrosion inhibitor
Oakite Products, Inc.	Enprox	nonchromated inhibitors for scale and corrosion

TABLE 5—*Selected commercially available alterants.*

Manufacturer	Alterant	Comment
FMC Corp.	hydrogen peroxide, oxygen	
Pentech Div. (Houdaille Industries, Inc.)	JAC oxyditch systems	aerator system
Permutit	air system	aeration
Wallace and Tiernan (Pennwalt Corp.)	chlorination	
Capital Controls Co. (Dart Industries)	chlorination	
IMC Chemical Group	AMP-95	neutralizing amine for CO_2
C-E Bauer (Combustion Engineering, Inc.)	airpact air diffusers	aerator system
Xodar Corp.	360 system	aerator
Rohm and Haas	ion exchange	metals removal
Various manufacturers	hydrochloric acid	reduced scale at Salton Sea
DuPont Co.	TYSUL WW, hydrogen peroxide	
Ionac Chemical Co.	ion exchange	
Dow Chemical Co.	ion exchange	

and 4.0 with HCl, scaling in the nozzles was eliminated and substantially reduced on the wear blades. Acidified brine effluents remained clear several hours after collection, whereas the unmodified brine was slightly turbid when collected, with precipitates forming a few minutes after samples had been taken [2].

Water Dilution

The addition of water to the fresh fluid was successful in reducing silica scaling at Namafjall, Iceland. Before dilution, scale was deposited from 95°C water as loose, leaflike flakes, which grew to 15 to 30 mm inside a 20.3-cm (8-in.) pipe. The scaling was reduced by mixing unflashed fluid from the drillhole to a 35 percent dilution with cold water at atmospheric pressure. The addition of dilution water reduced the silica content of the fresh fluid from 347 ppm to 188 ppm [2].

Before relying on dilution to reduce silica precipitation, one needs to consider that the diluent must be chemically compatible with the brine. For instance, attempts to reduce scaling in the GLEF at Niland, Calif., by the addition of steam condensate to the brine actually produced higher rates of scale and solids formation. This was a result of the high ammonia and carbonate content of the condensate and its correspondingly high pH (9 to 10). The problem is that when steam containing noncondensibles is cooled, a redistribution of species occurs with most of the ammonia redissolving. This raises the pH of the condensate and promotes dissolution of CO_2 into the condensate [2].

Oxidation and Aeration

The oxidation of sulfide to sulfur or sulfate—for example, by hydrogen peroxide, nitric acid—has been proposed as a means of controlling sulfide scale deposition [2]. A problem here is the cooling of the brine and the formation of insoluble metal sulfates [for example, calcium sulfate ($CaSO_4$)] and elemental sulfur, which cause erosion of the piping and plugging of the injection systems. The addition of a dispersing agent may be needed to prevent the solids from settling out.

In water quality treatment, diffused-air aerators are used to remove gases such as H_2S and CO_2 and to oxidize Fe^{++} and Mn^{++} to form precipitates. The method utilizes injection of compressed air through a perforated pipe or a similar system to produce fine bubbles. A portion of the CO_2 is exchanged from the water phase to the gas phase, as determined by Henry's law; the H_2S is oxidized. An advantage of aeration for H_2S removal is the low cost of the air used in aeration. However, aeration can cause formation of sulfate and the subsequent deposition of insoluble metal sulfates, as well as corrosion, because of the introduction of excess dissolved oxygen.

Coagulants

Coagulants are chemicals, such as aluminum sulfate, ferrous sulfate, and sodium aluminate, added to waters and wastewaters to speed up the settling and deposition of suspended solids. Flocculants to aid coagulation are often added. The method has been used to treat spent brines prior to disposal. See Table 6 for a list of typical coagulants.

The preferred method for disposal of geothermal waste brine is injection into subsurface formations to prevent mixing of the spent fluid with water that will be used for irrigation or drinking purposes. An important step to consider in waste brine injection is pretreatment to prevent precipitation or other reactions between the brine and the injection well system, which will cause scale, corrosion, or solids deposition.

Laboratory studies are needed for designing a treatment and injection disposal system. The required data include the following: (1) chemical composition of the brine, (2) pH, (3) scale-forming potential, (4) compatibility of the brine and water in the aquifer, and (5) necessary treatment for stabilizing the brine for injection into the desired subsurface formation [2,17]. The laboratory studies should provide data on the kind of treatment needed; an extensive treatment system could require the following steps:

(*a*) storage of the spent brine, in an open or closed storage container;

(*b*) corrosion control—pH adjustment, inhibitors;

(*c*) solids separation, by means of gravity, sedimentation, coagulation, or filtration;

(*d*) slime control, using a biocide or chlorine;

(e) oxidation by removing arsenic as pentavalent arsenic [As(V)], and oxidizing H_2S;
(f) aeration for gas/liquid exchange, removing CO_2; and
(g) ion exchange, adsorption—boron control.

A Specific Example: Treatment Methods for Removal of Boron from Geothermal Brines

The presence of boron in geothermal brines at concentrations ranging up to several hundred parts per million [18] represents a potential hazard from the environmental standpoint. Boron, in the form of boric acid and its water-soluble salts, is an essential trace element in the normal growth of all higher green plants [19]. However, as is the case with most trace-element nutrients, excessive amounts become toxic and lead to plant death. Some boron-sensitive plants and crops show toxic effects when irrigated with water containing boron concentrations as low as 1 ppm [20].

The development of economically feasible methods for removal of boron at the parts-per-million level from wastewaters—industrial, agricultural, and geothermal—is still in progress [21]. Various approaches have been employed with varying degrees of success, as outlined in the following paragraphs.

The adsorption of boric acid onto hydrated aluminum oxide and iron oxide sites in clays and clay components of soils could provide an effective method for boron removal under certain circumstances [21-24]. Unfortunately, the adsorptive capacities of clays and related materials are low (0.1 mg/g) even under optimum conditions. From the plant engineering standpoint,

TABLE 6—*Selected commercially available coagulants and flocculants.*

Manufacturer	Coagulant	Comment
Allied Colloids	Percol 726 (anionic high molecular weight polymer bead)	flocculant
Hercules, Inc.	Hercofloc organic polymers	flocculant for suspended solids
ARRO Laboratories, Inc.	ARRO-CX (cellulose xanthate)	removes heavy metals
Petrolite Corp.	Tolfloc 352, 300	flocculant, cationic polyelectrolyte
Nalco Chemical Co.	instant polymers	flocculant
Narvon Mining	Zeta Floc WA	polyelectrolyte flocculant
Swift Environmental Systems Co.	Lectro Clear (alum)	electrocoagulation
	Watcon 1355	high molecular weight cationic flocculant
Oakite Products, Inc.	Enprox	for coagulation
Dow Chemical Co.	Purifloc	

prohibitively large amounts of adsorbate would be required in most cases. However, adsorption processes are an important consideration in the removal of boron by soils *in situ* following a spill or other accidental contamination.

Precipitation methods commonly used in the water treatment industry, that is, those involving lime, aluminum hydroxide, and iron hydroxide, are ineffective in removing boron [23,25]. In addition, the conventional sedimentation and biological treatment processes employed in sewage treatment plants have little or no effect on boron levels [26].

Reverse osmosis (through cellulose acetate) would appear to have but limited application in the removal of boric acid from wastewaters. At an operating pressure of 3447.4 KPa (500 psi), only 20 to 30 percent of the boron in seawater is rejected, compared with over 95 percent of the sodium chloride [27,28].

Of the wastewater treatment processes, those involving ion-exchange resins appear to be the most promising [29,30]. It is recognized, of course, that the complete deionization of water by strongly basic ion-exchange resins such as Amberlite IRA-400 and Dowex 2 is not an economical approach for the removal of boron from waters containing appreciable concentrations of other exchangeable anions, such as the chloride ion. However, the strongly basic resins might be employed economically in the removal of boron from geothermal steam condensates that are relatively low in total salinity. Similarly, the Amberlite IRA-400 and Dowex 2 resins could be used in the removal of boron from brackish waters previously treated by one or more reverse-osmosis cycles.

However, it appears that the most direct and economical method for removal of boron from geothermal brines may be the use of the recently developed boron-specific resin, Amberlite XE-243 [24,30]. This is a crosslinked microreticular gel-type polymer derived from the amination of chloromethylated styrene-divinyl benzene with *N*-methylglucamine. The resin exhibits equilibrium boron capacities in the area of 5 mg/g with good hydrodynamic properties and chemical stability. The boron selectivity of Amberlite XE-243 is not affected by high concentrations of various other salts, including sodium chloride. It has been estimated that the overall costs for lowering boron concentrations from 10 to 1 ppm in irrigation waters would be below $0.03 per 3785 litres (1000 gal). An extensive study of the applicability of Amberlite XE-243 in the deboronation of geothermal brines appears warranted.

Corrosion

The treatment methods which are currently in use or which might be useful in controlling geothermal corrosion generally fall into one of two categories: (1) the removal of brine constituents that cause corrosion and (2) the selection or development of corrosion-resistant materials.

Corrosion rates in geothermal fluids that are complex solutions containing many chemical species are dependent primarily on the following most significant species: H^+, Cl^-, H_2S, CO_2, $CO_3^=$, HCO_3^-, ammonia (NH_3), and $SO_4^=$. Other less common or less aggressive chemical species that can also produce corrosive effects in some geothermal fluids are F^-, heavy metals, boron, and oxygen (O_2). In addition, the presence of heavy or transition metal ions can also be corrosive. See Table 3 [31].

Aeration, a process used in "open-type" systems, involves a mass transfer between the water and gas phases to speed up the removal of acidic gases (for example, H_2S and CO_2) by producing a large contact surface area between the air and water. Typical aeration equipment includes cooling towers, spray nozzles, and forced draft blowers.

Chemical degasification is used to remove oxygen selectively from the water by adding a chemical such as sodium sulfite or hydrazine to remove O_2 from oil-field brines and boiler feed water. Sodium sulfite (10 ppm Na_2SO_3 per 1 ppm O_2) was added to the 86°C water in the Reykjavik municipal heating system in Iceland to reduce oxygen and thereby control internal corrosion of metals in the heating systems [31].

Current methods of corrosion control are centered on planned replacement of turbine blades, piping, and other plant components and the development of materials with improved corrosion resistance. Table 7 [32,33] presents a list of materials tested for their performance in liquid-dominated geothermal resources. The reader is referred to Refs 32 and 34 for a thorough review of the corrosion resistance of various metals and alloys in hot brines.

Computer-Assisted Calculations of Scaling Rates

Computer-assisted calculations are finding increased use in the prediction of scaling and corrosion. A great advantage is the ability to simulate the effects of changes in brine variables, such as temperature and composition, and the probable result on scaling and corrosion (see Table 8).

The need is to estimate scaling and corrosion rates at various points in a power plant. Input data are based on the brine chemistry and flow, and the code provides the following: flow rate, temperature, and velocity at points from the bottom of the production well through the plant to the waste injection system [35,36,37].

Valid analytical expressions and supporting rate data are needed to calculate scaling. The general approach to scaling rate is to equate the rate of buildup to the degree of insolubility or supersaturation of a mineral minus the rate of mechanical removal. For example, the EQUILIB GEOTHERM computer code calculates the degree of mineral precipitation from the temperature, volume, composition, pH, and gas content of the brine [35].

TABLE 7—*Commonly used methods for controlling corrosion.*

Site	Method	Comment
Raft River, Well No. 1	materials evaluation: bronzes brasses, stainless steels, nickel-based alloys, cobalt-based alloys, and titanium had lowest corrosion rates	major brine constituents for corrosion: Cl, HCO_3, CO_3, S, CO_2, H_2S, Fe, Ca, Mg, SiO
Salton Sea, Magmamax No. 1	materials testing: low Cr-Mo steels had the lowest rate	
Desalination	deaeration, pH control, polyphosphate addition, materials (titanium, polymers)	chromate and phosphate mixtures promising
Oil-field fouling	antifoulants composed of mixture of dispersants, antioxidants, corrosion inhibitor	see Ref 33 for detailed discussion
	pre-Krete G-8 and C-17 blocks	available from Pocono Fabricators (Patterson-Kelley Co.)
	water-soluble molybdates	Climax Molybdenum Co.
East Mesa[a]	≤ 0.00254 cm/year (≤ 1 mils per year) corrosion for A1S1 Type 302, Type 430, Hastelloy S, titanium	data for wellhead fluid
Heber[a]	≤ 0.00254 cm/year (≤ 1 mils per year) corrosion for A1S1 316L, titanium, Hastelloy G, Inconel 625	data for wellhead fluid
Baca location No. 1[a]	materials testing: ≤ 0.00254 cm/year (≤ 1.0 mils per year) for A1S1 Type 316 Carpenter 20Cb3, Carpenter 7Mo, Inconel 600, Incoloy 825, titanium; >0.0254 cm/year (>10 mils per year) for carbon steel	data for flashed wellhead fluids

[a] See Ref 32 for additional data.

TABLE 8—*Selected computer codes for geothermal and other waters.*

Code	Comment
Helgeson-Herrick	predicts sulfide-silicate precipitation from Salton Sea brine
EQUILIB	predicts control of scale formation on adding acid, equilibrium scale formation
FLOSCAL	kinetics of scaling
PLANT	impact of scale buildup
GEOSCALE	identifies plant scale problems
BROP7	thermodynamic properties of brine and vapor
WELLFLOW	wellhead flow for geothermal well
CALGUARD	predicts corrosion and deposition, along with appropriate treatment in operating cooling systems

Summary and Conclusions

In summary, currently used methods of controlling scaling in geothermal plants are mainly centered on mechanical means, such as using wire brushes to remove scale from pipes and acidizing or reaming wells. Scale removal can be considered a maintenance problem and scheduled as needed.

Recommendations

A research and development program with the purpose of preventing the formation of scale might include the following:

1. Calcite formation in production wells can be controlled by maintaining the downhole pressure to prevent loss of CO_2 by flashing. Maintaining the pressure keeps CO_2 dissolved in the brine as bicarbonate and prevents the formation of insoluble calcium carbonate. The development of downhole pumps is currently in progress.

2. Test equipment is needed to provide empirical data on methods of controlling geothermal scale formation, for example, by evaluation of various commercial additives. The equipment should preferably be convenient for use at the laboratory bench and provide test results under simulated field conditions within a short time, about 2 to 3 days, using standard testing methods.

3. Fundamental studies are needed on the growth and inhibition of geothermal scales (for example, of FeS, SiO_2, $CaCO_3$). The studies should include both heterogeneous scale formation on solid surfaces, such as pipes and turbines, and homogeneous nucleation and formation of the main scaling substances within the geothermal brine. The results of some laboratory studies for silica and calcite formation are available.

4. Scale inhibitors specifically designed and developed for geothermal applications are needed. These should be tested for possible use at each location because brine composition and flow rates can vary from site to site. A program is in progress to evaluate commercially available scale inhibitors that have been developed mainly for nongeothermal fluids.

5. Studies on the adhesion of geothermal scales could provide information on methods to prevent scales from depositing onto pipes, turbines, instruments, and other equipment.

6. A data base containing data on the rate of formation of scale substances, such as antimony sulfide (Sb_2S_3), arsenic sulfide (As_4S_3), $CaCO_3$, SiO_2, FeS, would be valuable for predicting the likelihood of scale formation. The data are needed for incorporation into existing computer data bases, which currently contain mainly equilibrium values.

7. Instrumentation is needed to monitor geothermal plant parameters to measure scale buildup. The instruments should be rugged and able to function in a high-temperature environment; self-cleaning features to remove

scale or the ability to measure without being in contact with the brine are desirable features. Instruments of this type are available for monitoring wastewaters at low temperatures.

8. Treatment methods are needed to remove from waste fluids selected constituents, such as boron, which are not removed by conventional methods, for example, by holding tanks and coagulants, and to hasten the precipitation of silica. In the latter case, seeding with brine solids is being studied.

Acknowledgment

Thanks are extended to Daniel A. McLean, East Bay Municipal Utility District, Oakland, Calif., and to Oleh Weres, Lawrence Berkeley Laboratory, Berkeley, Calif., for their comments. This report was done with support from the U.S. Department of Energy, under Contract No. W-7405-ENG-48.

References

[1] Lyon, R. N. and Kolstad, G. A., "A Recommended Research Program in Geothermal Chemistry," WASH-1344, U.S. Department of Energy, National Technical Information Service, Springfield, Va., 1974.
[2] Phillips, S. L., Mathur, A. K., and Doebler, R. A., "A Study of Brine Treatment," EPRI ER-476, Lawrence Berkeley Laboratory, Berkeley, Calif., 1977.
[3] Reid, G. W., Streebin, L. E., Canter, L. W., and Smith, J. R., "Brine Disposal Treatment Practices Relating to the Oil Production Industry," EPA-660/2-74-037, University of Oklahoma Research Institute, Norman, Okla., 1974.
[4] Water Quality and Treatment, A Handbook of Public Water Supplies, 3rd ed., American Water Works Association, McGraw-Hill, New York, 1971.
[5] Vetter, O. J., Journal of Petroleum Technology, 1972, pp. 997–1006.
[6] Collins, A. G., International Symposium on Oilfield and Geothermal Chemistry, Paper No. SPE 6603, La Jolla, Calif., 1977, pp. 37–44.
[7] Harrar, J. E., Lorensen, L. E., Otto, C. H., Jr., Deutscher, S. B., and Tardiff, G. E., Annual Meeting of the Geothermal Resources Council, Hilo, Hawaii, 1978, Vol. 2, pp. 259–262.
[8] Harrar, J. E., Locke, F. E., Lorensen, L. E., Otto, C. H., Deutscher, S. B., Rey, W. P., and Lim, R., "On-Line Tests of Organic Additives for the Inhibition of the Precipitation of Silica from Hypersaline Geothermal Brine," UCID-18091, Lawrence Livermore Laboratory, Livermore, Calif., 1979.
[9] Chernozubov, V. B., Zaostrovskii, F. P., Shatsillo, V. G., Golub, S. I., Novidov, E. P., and Tkach, V. I., Proceedings, First International Symposium on Water Desalination, Washington, D.C., 1965, Vol. 1, pp. 539–547.
[10] Harrar, J. E., Locke, F. E., Otto, C. H., Jr., Deutscher, S. B., Lim, R., Frey, W. P., Quong, R., and Lorensen, L. E., "Preliminary Results of Tests of Proprietary Chemical Additives, Seeding, and Other Approaches for the Reduction of Scale in Hypersaline Geothermal Systems," Report No. UCID-18051, Lawrence Livermore Laboratory, Livermore, Calif., 1979.
[11] Lindemuth, T. E., Houle, E. H., Suemoto, S. H., and Van Der Mast, V. C., International Symposium on Oilfield and Geothermal Chemistry, Paper No. SPE 6605, La Jolla, Calif., 1977, pp. 173–186.
[12] Tomson, M. B., Nancollas, G. H., and Kazmierczak, T. E., International Symposium on

Oilfield and Geothermal Chemistry, Paper No. SPE 6591, La Jolla, Calif., 1977, pp. 13-20.
[13] DeBoer, R. B., *American Journal of Science,* Vol. 277, 1977, pp. 38-60.
[14] Metcalf, J. H., National Association of Corrosion Engineers, Houston, Tex., 1974, pp. 196-219.
[15] Hausler, R. H., *Oil and Gas Journal,* 1978, pp. 146-154.
[16] Michels, D. E. and Keiser, D. D., Conference on Scale Management in Geothermal Energy Development, Report No. C00-2607-4, San Diego, Calif., 1976, pp. 51-58.
[17] *Salt Water Disposal, East Texas Oil Field,* 2nd ed., East Texas Salt Water Disposal Co., Kilgore and Tyler, Tex., 1958. Available from the Petroleum Extension Service, the University of Texas, Austin, Tex.
[18] Bassett, R. L., "The Geochemistry of Boron in Geothermal Waters," Ph.D. thesis, Stanford University, Stanford, Calif., 1976.
[19] Bonner, J. E. and Varner, J. E., "Plant Biochemistry," Academic Press, New York, 1976.
[20] Bingham, F. T., *Advances in Chemistry Series,* Vol. 123, 1973, p. 130.
[21] Choi, W. W. and Chen, K. Y., *Environmental Science and Technology,* Vol. 13, 1979, pp. 189-196.
[22] Sims, J. R. and Bingham, F. T., *Proceedings,* Soil Science Society of America, Vol. 31, 1967, p. 728.
[23] Jasmund, K. and Linder, B., "Experiments on the Fixation of Boron by Clay Minerals," Proceedings, International Clay Conference, Madrid, Spain, 1972.
[24] "Chemical Technology and Economics in Environmental Perspectives. Task II—Removal of Boron from Wastewater," Report No. EPA-560/1-76-007, U.S. Environmental Protection Agency, Office of Toxic Substances, Washington, D.C., 1976.
[25] Waggott, A., *Water Research,* Vol. 3, 1969, p. 749.
[26] Roberts, R. M. and Gressingh, L. E., "Development of Economical Methods of Boron Removal from Irrigation Return Waters," Report No. 579, U.S. Department of the Interior Research and Development, Washington, D.C., 1970.
[27] Lonsdale, H. K., "Study of Rejection of Various Solutions by Reverse Osmosis Membranes," Report No. 447, U.S. Department of the Interior Research and Development, Washington, D.C., 1969.
[28] Cruver, J. E., *Water and Sewage Works,* Vol. 13, 1973, p. 74.
[29] Grinstead, R. R. and Wheaton, R. M., "Improved Resins for Removal of Boron from Saline Waters," Report No. 721, U.S. Department of the Interior Research and Development, Washington, D.C., 1971.
[30] Kunin, R., *Advances in Chemistry Series,* Vol. 123, 1973, p. 139.
[31] Hermannsson, S., *Geothermics,* Vol. 2, Special Issue No. 2, 1970, pp. 1602-1612.
[32] DeBerry, D. W., Ellis, P. F., and Thomas, C. C., *Materials Selection Guidelines for Geothermal Power Systems,* 1st ed., Report No. ALO-3904-1, Radian Corp., Austin, Tex., 1978.
[33] Nathan, C. C., "Corrosion Inhibitors," National Association of Corrosion Engineers, Houston, Tex., 1974.
[34] Banning, L. H. and Oden, L. L., "Corrosion Resistance of Metals in Hot Brines—A Literature Review," Report No. 1C-8601, U.S. Bureau of Mines, Washington, D.C., 1973.
[35] Shannon, D. W., Walter, R. A., and Lessor, D. L., "Brine Chemistry and Combined Heat/Mass Transfer," Report No. EPRI ER-635, Battelle Pacific Northwest Laboratories, Richland, Wash., Vols. 1 and 2, 1978.
[36] *Industrial Water Engineering,* Vol. 15, 1978, p. 4.
[37] Riney, T. D. and Reynolds, S. L., *Proceedings,* Report No. EPRI WS-78-97, Second Geothermal Conference and Workshop, Taos, N. Mex., 1978, pp. 147-150.

R. E. McAtee,[1] *C. A. Allen,*[2] *and L. C. Lewis*[3]

Chemical Logging of Geothermal Wells

REFERENCE: McAtee, R. E., Allen, C. A., and Lewis, L. C., **"Chemical Logging of Geothermal Wells,"** *Geothermal Scaling and Corrosion, ASTM STP 717,* L. A. Casper and T. R. Pinchback, Eds., American Society for Testing and Materials, 1980, pp. 225-235.

ABSTRACT: Chemical logging was developed to permit identification of zones bearing geothermal water during drilling of geothermal wells. This technique consists of collection of circulation fluid samples at specified periods determined by drill-bit depth. These samples are analyzed for bicarbonate (HCO_3^-), Cl^-, F^-, Ca^{++}, and silicon dioxide (SiO_2). The results are plotted as a function of depth, and the resulting log indicates the presence of water-bearing structures. It was discovered that the ratio of Ca^{++} and HCO_3^- concentrations is a temperature indicator. It was also found that chemical composition changes precede structural changes. This report describes chemical logging of three wells at the Raft River geothermal site, Idaho.

KEY WORDS: chemical logging, geothermal, logging, geothermal drilling, scaling, corrosion

The growth of the geothermal industry has created a need for techniques to be used during drilling operations that will assist in determining well depth, casing location, and well development. Geophysical logging, lithologic logging, and core drilling are important techniques developed by the petroleum industry. Other exploration and reservoir evaluation techniques, developed by the petroleum industry, universities, and government agencies, are useful. However, very little development has been specific to geothermal well drilling and exploration. The chemical logging method described in this report was developed for geothermal application. It is a technique carried out during drilling operations.

The chemical logging studies were conducted at the Raft River geothermal site in south-central Idaho. Seven geothermal wells have been drilled to sup-

---

[1] Senior chemist, E G & G Idaho, Inc., Idaho Falls, Idaho 83401.
[2] Manager, Biological and Earth Sciences, E G & G Idaho, Inc., Idaho Falls, Idaho 83415.
[3] Manager, Analytical Services, Exxon Nuclear Idaho Co., Inc., Idaho Falls, Idaho 83415.

port experiments and to run a binary cycle power plant. It is a U.S. Department of Energy experimental site. Studies are being conducted on binary heat exchangers, aquaculture, agriculture, and corrosion, utilizing the geothermal water.

Chemical logging began during the drilling of Well RRGE-3 at the geothermal site. Samples of the drill fluid were collected for chemical analysis periodically during drilling. Chemical analysis was performed to determine the concentration of chemical geothermometer species and other indications of geothermal water. The test results were significant enough to warrant planning a more complete test on the next well-drilling operation scheduled at the Raft River geothermal site.

The sampling of the drill return fluid for the determination of certain borehole conditions is not unique. Lithologic logging is routine for most major drilling operations. Drill water, inlet and outlet temperatures, pH, flow rates, methane, and hydrogen sulfide are the parameters most often measured on the drill fluid. The most important phase of the operation is the geological evaluation of the drill cuttings to study the borehole lithology. Elders, Hoagland, and McDowell [1][4] used drill cuttings to determine the hydrothermal alterations in borehole geology; with this information, they predicted the observed temperature of a completed geothermal well. These applications could be considered geochemical logging. However, prior to this study, the drill cuttings were analyzed—not the drill water.

This report describes the chemical logging study of three geothermal wells. Chemical logs are prepared by analyzing the drill return fluid and graphically plotting the concentrations of the analyzed chemical species to the drill depth. It is an empirical technique, and few theoretical studies have been made at this time.

Changes in the chemical composition of the drill fluid result from dilution by the drill mud and the water-bearing strata penetrated by the drill string. Figure 1 is a cross-sectional view of a drill string that has penetrated several water-bearing strata. Drilling fluid is pumped through the drill stem and bit. The fluid then returns to the surface between the drill stem and the wall of the borehole, carrying the drill cuttings with it.

When a water-bearing stratum is penetrated, as in this illustration, it contributes water to the drilling fluid. This dilutes the drilling fluid and causes variation in its chemical composition. Generally, after the drill string passes through the water-bearing stratum, the drilling mud or sediments in the drill water form a mud cake on the walls of the borehole. However, if the flow from the stratum is large and the mud cake does not seal it, the incoming water will change the chemical background of the drilling fluid. The determination of the change in chemical composition of the diluted drill fluid and

---

[4]The italic numbers in brackets refer to the list of references appended to this paper.

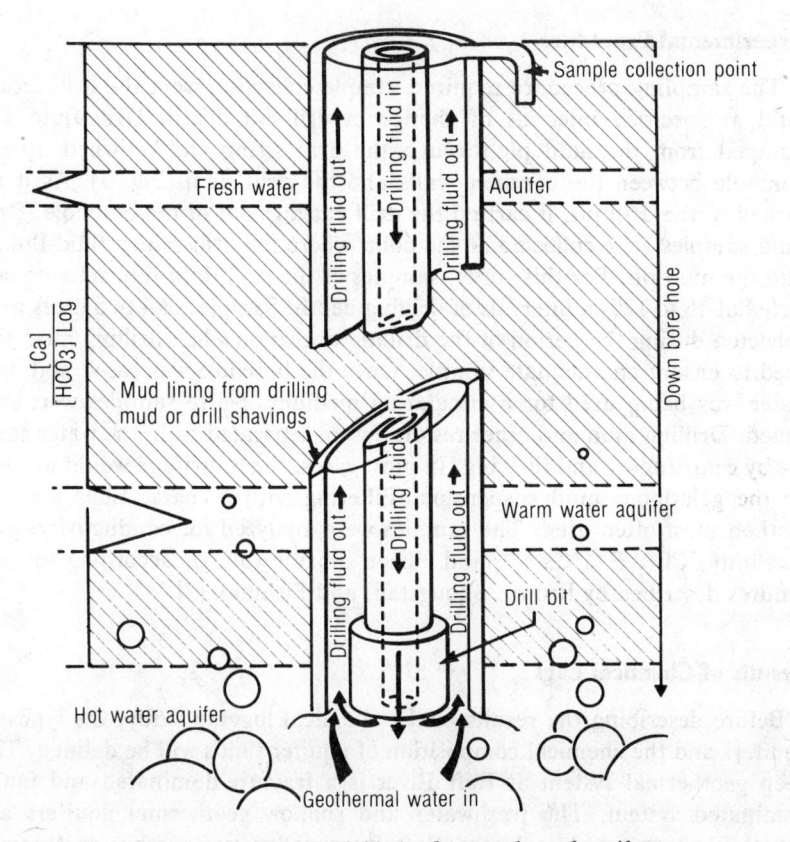

FIG. 1—*Drilling fluid dilution by a geothermal aquifer.*

its separation from the chemical background contributed by the drill fluid, drill mud, and other aquifer leakage is the essence of the chemical log.

The ratios of chemical species found to be useful in geochemical evaluations by other researchers [2] were used to develop chemical logs.[5] At present these logs are being correlated to the geophysical logs. The correlations have already revealed that the ratio of the concentrations of calcium and bicarbonate ions can be used as a relative temperature indicator.

Other applications of chemical logs are being studied. Correlation between the logs of the ratios of chemical species and the geophysical logs are being evaluated. It is evident that chemical logging can be a useful tool in the drilling of geothermal wells and the exploration of geothermal resources.

---

[5] Overton, H. L., senior staff engineer, Cer Corp., Las Vegas, Nev., personal communication.

Experimental Procedure

The sampling procedure required sample collection from the drill return fluid at specified intervals of change in drill bit depth. Drill fluid was pumped from the mud pit through the drill string and returned up the borehole between the borehole walls and the drill stem (Fig. 1). As it returned to the drill pit, it carried the drill cuttings out of the borehole. Drill fluid samples were collected at the point where the drill return fluid flowed into the mud pit. For this study, samples of the drill fluid returns were collected at 15 to 120-m intervals of drilling depth. Samples of 4 to 5 litres were collected during the period of the drilling operation when drilling mud was used to ensure an adequate sample. Once the borehole was cased and only water was being used for a circulating medium, 1-litre samples were collected. Drilling mud and other residues were separated from the water sample by centrifuging and filtering. In most cases, the centrifuge would not settle the gelatinous mud suspension. Filtering with a coarse filter was the method most often used. The samples were analyzed for conductivity, pH, alkalinity, Cl^-, F^-, Ca^{++}, and silicon dioxide (SiO_2), according to procedures described by Brown, Skougstad, and Fishman [3].

Results of Chemical Logs

Before describing the results of the chemical logging study, the types of aquifers and the chemical composition of aquifer fluids will be defined. The deep geothermal system at Raft River is a fracture-dominated and fault-dominated system. The freshwater and shallow geothermal aquifers are permeable systems. Also in the Raft River valley, as in other geothermal areas, freshwater aquifers are affected by geothermal water intrusion. Table 1 shows the effect of these variables on the chemical composition of the drilling fluid. Geothermal water was used for drilling fluid in this example, because most of the geothermal wells at the Raft River geothermal site were drilled with geothermal water as the drilling fluid.

TABLE 1—*Variables to the chemical composition of geothermal water used for drilling fluid.*[a]

Variable	$[Ca^{++}]/[HCO_3^-]$	F^-	Cl^-	Ca^{++}	SiO_2	pH	Conductivity	Alkalinity
Drilling mud	ND	—	—	—	—	+	—	+
Geothermal water intrusion	+	+	+	+	+	ND	+	—
Freshwater intrusion	—	—	—	+	—	ND	—	+
Geothermal contaminated freshwater intrusion	+	ND	ND	ND	ND	ND	ND	—

[a]The symbols used are defined as follows:
   + = increased concentration of this species in the drill water.
   − = decreased concentration of this species in the drill water.
ND = not definable.

When the drill mud is added to drill fluid, a decrease is observed in the concentration of most of the analyzed chemical species. Increases in alkalinity and pH also result from adding drill mud to the drill fluid. Chemical composition changes from geothermal water intrusion are not definable, except for the increase in the calcium/bicarbonate (Ca/HCO$_3$) ratio and the decrease in alkalinity. Freshwater intrusion into the drill water would generally increase hardness and alkalinity and decrease F$^-$, Cl$^-$, conductivity, and the calcium/bicarbonate ratio. A geothermal and freshwater mixture intrusion into the drill water would result in increases in the calcium/bicarbonate ratio. The changes in the other chemical species would not be definable. Table 1 can be used as a guideline for the interpretation of the chemical logs.

A chemical log can be prepared by plottting graphically the concentrations of the analyzed chemical species to the drill string depth. The resulting log is a profile of the chemical change taking place in the drill fluid during the drilling operation. This information is very useful in evaluating the water-bearing stratum penetrated by the drill string.

Figure 2 is an example of this type of log. Evaluation of this log shows a freshwater aquifer's influence on the chemical composition of the drill fluid. At a depth of 488 m there are sharp increases in the alkalinity and hardness and sharp decreases in the F$^-$ and SiO$_2$ concentrations. This would be a typical change in chemical composition when freshwater dilutes the drilling water. The increase in chloride ion concentration and conductivity also indicate, however, that the freshwater aquifer had geothermal water intrusion. The production zone of a geothermal aquifer was penetrated at 1280 m. Because the drill fluid was geothermal water similar to that in the aquifer, only small changes were observed in the drill fluid chemical composition. There was a small increase in the SiO$_2$ concentration and a small decrease in conductivity. The decrease in the alkalinity concentration was the only large change detected at this depth. However, these and many other interpretations could be made confidently from this type of chemical log. Concentration changes in geothermometers, such as changes in the SiO$_2$, can be used to determine relative aquifer temperatures. Conductivity can be used as an indicator of water quality. All of these parameters can be determined during the well-drilling operation.

Results of Calcium/Bicarbonate Chemical Logs

Chemical logs of the ratios of the analyzed chemical species were evaluated. Chemical ratios used to indicate hydrothermal alteration and geothermometers were studied. It was found that taking ratios of concentrations of calcium and bicarbonate ions and plotting them graphically to depth resulted in a chemical log similar to the temperature log of a geothermal well. This study also showed that as the drill approached the geothermal zone, the

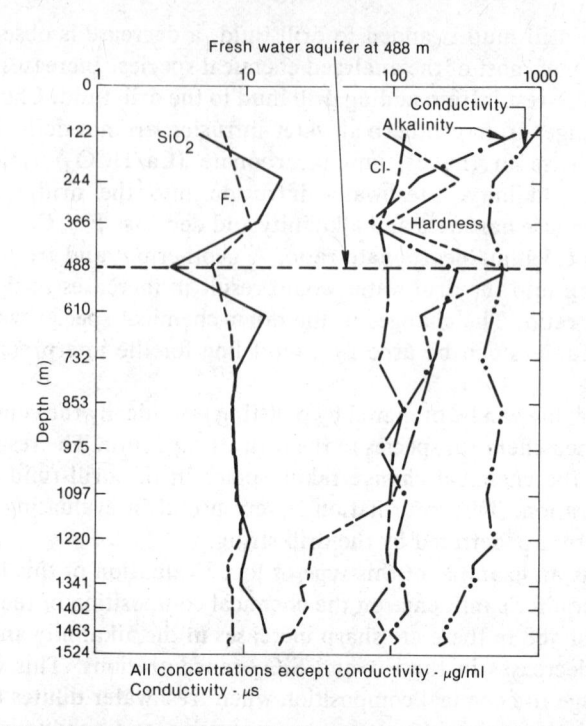

FIG. 2—*Chemical log of all analyzed chemical species for Well RRGP-5.*

calcium/bicarbonate ratio increased, displacing the resultant log uphole in relation to the temperature log. This displacement appeared to be due to leakage or diffusion of hot water from the aquifer. The uphole displacement varied between 16 and 120 m for the wells tested. Displacement appears to be a function of the permeability of fractured material above the geothermal aquifer. This characteristic of the calcium/bicarbonate ratio for anticipating when the drill is approaching a geothermal aquifer, combined with the information on the permeability of the stratum already penetrated furnished by the mud logger, could be useful in determining the depth of the well casing.

The first program designed specifically for chemical logging of a well was conducted on Well RRGI-6 at Raft River. The purpose of the program was to characterize chemically the aquifers penetrated by the drill string. Samples were collected at every 120-m interval of depth to 911 m, and then at every 60 m to the total depth. When the well was completed, studies were made of the chemical and geophysical logs. Comparison of the calcium/bicarbonate ratio to the temperature log revealed the similarities shown in Fig. 3. The comparison also shows about a 60-m uphole displacement of the calcium/bicarbonate log in relation to the temperature log. This means the chemical log anticipated the geothermal zone 60 m before the drill penetrated it.

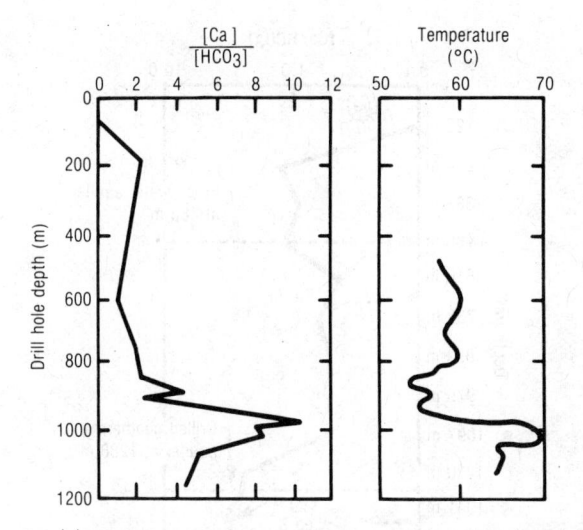

FIG. 3—$[Ca^{++}]/[HCO_3^-]$ chemical log and temperature log for Well RRGI-6.

The evaluation of chemical logs on Well RRGI-6 indicated a need for further verification. This was accomplished by chemically logging the next well drilled at the Raft River site (RRGP-5). Also, comparison of chemical logs to the geophysical logs indicated the need for increasing the number of samples to increase the resolution of the chemical log.

The following procedure was used for the next well-drilling operation:

1. The sample intervals were 120 m until changes were observed in the chemical log, indicating increasing temperature.
2. When increasing temperature was detected, the sampling interval was decreased to 60 m.
3. Samples were collected whenever the driller detected structural changes during the drilling operation.

The resulting chemical log, in which all the chemical species analyzed were plotted, is shown in Fig. 2. The calcium/bicarbonate chemical log is shown in Fig. 4. Evaluation of the calcium/bicarbonate log reveals a sharp increase in temperature at a depth of 1220 m. This increase in the calcium/bicarbonate ratio was observed until the drill string reached a depth of 1280 m. At this depth, a high flow of hot water was observed. The flow rate was estimated to be approximately 4100 litres/m. This caused the original drill water to become so diluted with water from the aquifer that only small chemical changes were observed in the chemical logs. Geothermal water from this depth to the total depth washed away the chemical profile of the well. The lower part of this borehole was lost when a concrete plug was set at the depth of 1051 m to install the well casing. After the well was cased and

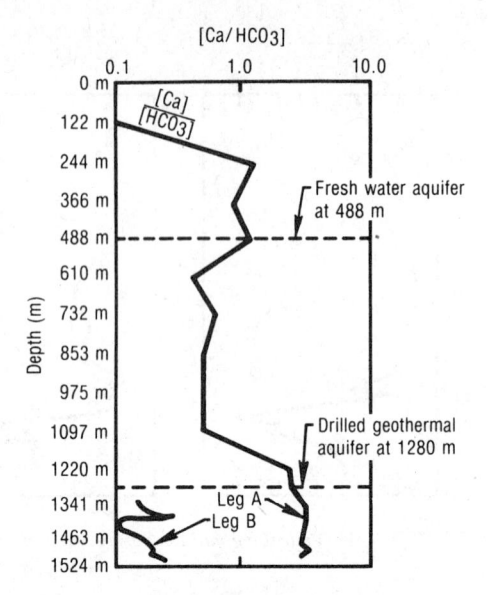

FIG. 4—$[Ca^{++}]/[HCO_3^-]$ chemical log for Well RRGP-5.

reentry was made with the drill string, it was not possible to drill through the concrete plug. Because of this, little geophysical logging was accomplished. Sidetrack drilling was initiated at the top of the plug. Either the second leg did not penetrate the high-flow zone penetrated in the first leg or the fractures were sealed with concrete. The second leg is shown as Leg B in Fig. 4. It is notable that the chemical log of Leg B does indicate the penetration of a narrow, hot-water-bearing aquifer, which later flowed at about 800 litres/m with a maximum temperature of 123°C.

Another variation in the calcium/bicarbonate log pertinent to the fluids' definition of the borehole was observed at the depth of 488 m. The presence of a freshwater aquifer at this depth was discussed previously. The increase of the calcium/bicarbonate ratio at this depth is an indication that this water was mixed with geothermal water

Figure 5 shows the calcium/bicarbonate chemical log of the redrilling of Well RRGI-4. The object was to convert an injection well into a production well. Sidetrack drilling started at 565 m, and the total depth of Leg A was 1650 m. To improve resolution, samples were taken at 15-m intervals, except in areas where the driller detected structural change; at these points, the interval was reduced to 8 m. To evaluate the chemical log as a rapid turn-around time tool, the calcium/bicarbonate chemical log was kept current with the drilling progress. The object was to detect any significant temperature changes to determine whether the drill string was approaching a production zone. At a depth of 1520 m the drill bit encountered a hard stratum prior to

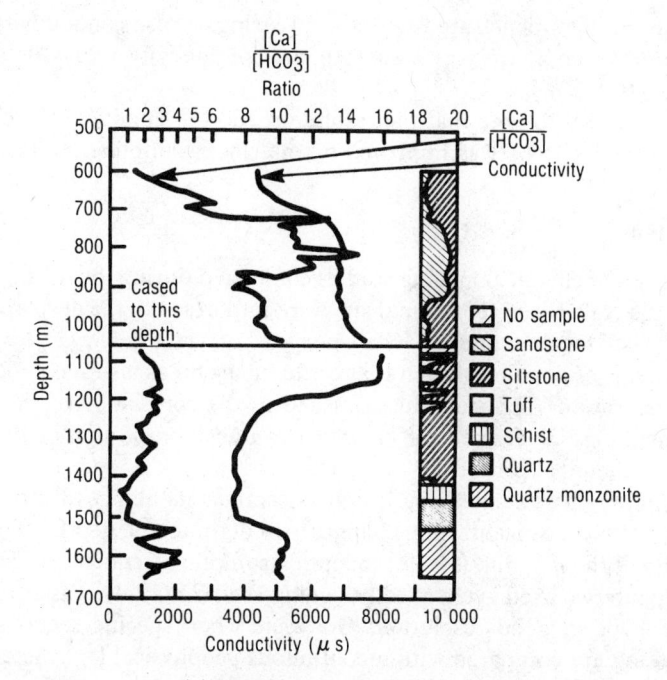

FIG. 5—$[Ca^{++}]/[HCO_3^-]$ chemical log for Well RRGP-4A.

the penetration of a narrow, low-producing, hot-water zone. The calcium/bicarbonate ratio began to increase when the drill reached the hard stratum. It continued to increase as the zone was penetrated, and it decreased after the drill passed through it. The driller noticed a change in drilling rate approximately 10 m before the drill penetrated the geothermal zone. This situation was repeated at 1580 m. The combined flow of the two production zones was about 135 litres/m, with water temperatures above the boiling point.

The drill penetrated another geothermal zone at a depth of between 700 and 870 m. The calcium/bicarbonate ratio increased sharply through this area. This interpretation is notable for two reasons. First, drilling mud was being used for drilling at this time, so drill fluid had to be separated from drilling mud to be analyzed. The background chemical composition of drill fluid is changed by the presence of the drilling mud. This can be seen in Fig. 5 by comparing the calcium/bicarbonate ratio before and after casing. Second, a comparison was made of the chemical log and the lithologic log. The section of the log where high calcium/bicarbonate ratios were experienced (between 700 and 870 m) corresponds very well with the increase in sandstone shown on the lithologic log. This sandstone stratum was verified to be an aquifer by geophysical logging.

The conductivity is also plotted on the log in Fig. 5 to show the water quality of the aquifers penetrated by the drill string. These conductivity values were found to compare very well with the conductivity values determined from flow tests.

It is important to note that the resolution and the quality of the chemical log of Well RRGP-4A was improved by the closer sampling intervals.

Conclusion

The series of chemical logging studies conducted during the drilling operations at the Raft River geothermal site were important to the development of the chemical logging method. The purpose of the study was to develop chemical logs of good resolution that could be useful as a support tool to the drilling operation. The study also assisted in the correlation of the various geophysical logs. It defined the chemical composition of aquifer fluids and developed chemical profiles of the borehole.

The quality of a chemical log is defined here as its ability to correlate accurately chemical concentration changes with definite well depths. This is the resolution of the log, and it is dependent on sampling frequencies. The 16-m sampling interval used in chemically logging Well RRGP-4A was adequate to produce a log of good resolution. However, when specific sections of the chemical log are compared with a continuous geophysical log, the chemical log is difficult to correlate if the comparable section of the geophysical log is small. This problem is being considered, and some potential solutions will be studied in future drilling operations.

The ability of the chemical log to furnish support to the drilling operation was evident during the drilling of Well RRGP-4A. The calcium/bicarbonate chemical log was used to determine whether the drill was approaching a geothermal production zone. Although the production of the well was small, the chemical log did indicate the two small, hot-water-bearing zones prior to penetration by the drill.

Planned work includes chemical logging of Well INEL-1 at the Idaho National Engineering Laboratory, Idaho Falls, Idaho, and two other resource areas. These experiments will determine the site-specific nature of chemical logging and will increase the scientist's ability to interpret logs from other resource areas. Analytical and log preparation techniques are being refined continually. This will improve interpretation of chemical logs and their correlation to geophysical and lithologic logs.

Acknowledgments

This study was supported by the U.S. Department of Energy, Division of Geothermal Energy, and is under the direction of E G & G Idaho, Geothermal Technical Division of the Idaho National Engineering Laboratory, Idaho Falls.

References

[1] Elders, W. A., Hoagland, J. R., and McDowell, S. D., "Hydrothermal Mineral Zones in the Geothermal Reservoir of Cerro Prieto, Baja California, Mexico," Abstracts, First Symposium on the Cerro Prieto Geothermal Field, Baja California, Mexico, 20-22 Sept. 1978, San Diego, Calif.

[2] Ellis, A. J. and Mahon, W. A. J., *Chemistry and Geothermal Systems*, Academic Press, New York, 1977, pp. 144-152.

[3] Brown, Eugene, Skougstad, M. W., and Fishman, M. J., *Techniques of Water-Resources Investigations of the USGS*, Book 5, Washington, D.C. 1970, Chapter A1, Part 4.

J. C. Watson[1]

Round-Robin Evaluation of Methods for Analysis of Geothermal Brine

REFERENCE: Watson, J. C., "**Round-Robin Evaluation of Methods for Analysis of Geothermal Brine,**" *Geothermal Scaling and Corrosion, ASTM STP 717,* L. A. Casper and T. R. Pinchback, Eds., American Society for Testing and Materials, 1980, pp. 236-258.

ABSTRACT: Geothermal liquid and gas analysis provides necessary information for evaluating corrosion and scaling, optimizing power plant design, assessing environmental impacts, and ensuring compliance with legal requirements. A basic problem with geothermal analysis is a lack of standardization of analytical methods.
  In 1976, Pacific Northwest Laboratory (PNL), operated by Battelle Memorial Institute for the U.S. Department of Energy, initiated a program under the U.S. Department of Energy–Division of Geothermal Energy (DOE-DGE) to determine the state of the art of methods of analysis for geothermal fluids. Following the issuance in 1976 of a comment manual [1][2] of literature-abstracted methods, an organizational meeting held in early 1977 led to the formulation of a round-robin field testing program. Attendees representing two government agencies, four government-funded laboratories, five private laboratories, and three industrial concerns agreed on 40 parameters to be evaluated. Two round-robins were conducted with approximately 20 laboratories participating in the analysis: the first sample collection, at the U.S. Bureau of Reclamation site in East Mesa, Calif., involved the evaluation of a low-solids (0.4 percent by weight) brine, while the second, conducted at the Geothermal Loop Experimental Facility (GLEF) of the San Diego Gas and Electric Co., Niland, Calif., involved a high-solids (24 percent by weight) brine.
  Statistical analysis of the data from these tests shows that the greatest range of results was obtained for the species silver, aluminum, antimony, fluoride, bromide, iodide, phosphate (PO$_4$), and total hydrogen sulfide (H$_2$S). In addition, the constituents rubidium, copper, manganese, lead, and zinc showed a high degree of variability for the brine with low total dissolved solids (TDS). Barium, bicarbonate (HCO$_3$), boron, potassium, sulfate (SO$_4$), arsenic, and total carbon dioxide (CO$_2$) were more difficult to analyze in the higher TDS brine than in the lower one.

KEY WORDS: water analysis, geochemistry, groundwater, water chemistry, chemical tests, chemical analysis, geothermal, scaling, corrosion

---

[1] Senior research scientist, Battelle, Pacific Northwest Laboratory, Richland, Wash. 99352; present position: supervisory chemist, U.S. Navy, Trident Refit Facility, Naval Submarine Base, Bangor, Bremerton, Wash. 98315.
[2] The italic numbers in brackets refer to the list of references appended to this paper.

The chemical analysis of geothermal fluids and gases is being done today by many investigators using their own choice of methods.

Analyses of geothermal samples are complicated by high total dissolved solids (TDS), matrix effects, and chemical interferences. Thus, the analytical numbers reported may not be truly representative of the geothermal systems.

The logical approach to the solution of these problems is to involve organizations with experience in geothermal analysis in assessing the state of the art, selecting candidate methods, verifying accuracy and reliability through "round-robin" testing, and to assess chemical interferences through systematic study of the methods. Out of this effort will evolve recommended procedures with guidance as to the applicability and the use of the methods.

Numerous interlaboratory comparison studies have been reported involving analysis of synthetic, freshwater, and seawater samples [2-12], however, few intercomparison studies involving geothermal brine samples have been undertaken [13]. This study is concerned with interlaboratory comparison studies of two geothermal brine samples for the purpose of evaluating state-of-the-art analytical methods. Present methodology consists primarily of standard freshwater and seawater analysis procedures, in accordance with such publications as the *1980 Annual Book of ASTM Standards, Part 31, Water* and other sources [14], with some modifications to compensate for the higher TDS, matrix effects, and chemical interferences attributable to geothermal brines.

The laboratories that participated in the Geothermal Round-Robin (GRR) Program (listed in the Appendix) included organizations concerned with geothermal investigations, representing industry, commercial analytical services, and government. Eighteen laboratories participated in the first round-robin study, which involved analysis of a low-TDS (0.4 percent) brine. An additional three laboratories participated in the second round-robin evaluation of the higher-TDS (24 percent) brine.

For these two studies, control samples were made up and supplied by Pacific Northwest Laboratory (PNL), Richland, Wash., to each of the participating laboratories to evaluate the consistency of interlaboratory analytical results. These control samples were prepared to simulate geothermal brines and contained nearly all of the chemical constituents given in Table 1. Parameters not included in the control samples were suspended solids, turbidity, bicarbonate, carbonate, and radioactive isotopes. All the participating laboratories received identical control samples. Numbers were assigned to all participating laboratories in order to conceal their identities. For the first round-robin test, the participating laboratories were supplied with a description of the sampling location, approximate concentrations of the brine components, and approximate concentrations of the control sample components. The only information supplied to the participating laboratories for the second round robin was a description of the sampling location. The brine and control samples were mailed directly from field locations to the participants to maintain consistency. The laboratories performed the evaluations and reported results, methods used, precision, accuracy, and comments to PNL.

Data of individual laboratory reports were tabulated for purposes of statistically evaluating the results. The data for both the control and brine samples were grouped according to the method of analysis used. The statistical terminology used here is that defined in the 1976 edition of the *Compilation of ASTM Standard Definitions* (American Society for Testing and Materials, Philadelphia). For each species, an average value, $\bar{X}$, the standard deviation, σ, the percentage coefficient of variation, percent CV, 95 percent confidence intervals about the average, and the acceptable range of results were computed. The statistical data are based on the 3σ rule-of-thumb criterion for rejection of data, that is, the average and standard deviation for each data set were calculated with the exclusion of values exceeding $\pm 3\sigma$ from the mean. This treatment is in accordance with the ASTM Recommended Practice for Dealing with Outlying Observations (E 178-75).

Results of these investigations were used to assemble a manual [15] of state-of-the-art procedures for sampling and analyzing geothermal fluids and gases.

Experimental Procedure

Parameters Evaluated and Sample Types

An organizational meeting was held in April 1977 with potential round-robin participants. The 14 participant organizations represented (see Appendix) agreed on a list of 40 parameters, shown in Table 1, to be evaluated for the two round-robin samples. Pacific Northwest Laboratory prepared the control samples, collected the samples at the two sites, and distributed the brine and control samples to the participants.

Samples at the site were collected in 49.2 litres (13-gal) polyethylene carboys, the contents of which were then split for distribution. The acidified samples were collected in carboys in which Ultrex[3] grade hydrochloric acid had been placed. Enough acid was added to yield a final acid concentration of 10 ml of acid per litre of brine. The total carbon dioxide (CO_2) and total hydrogen sulfide (H_2S) samples were collected utilizing the apparatus shown in Fig. 1. To collect carbon dioxide, brine was dispersed through a glass frit into a 2 N sodium hydroxide-absorbing solution contained in a 49.2-litre (13-

---

[3] Reference to a company or product by name does not imply approval or recommendation of the product by the U.S. Department of Energy or Battelle, Pacific Northwest Laboratory, to the exclusion of others that may have the same or better properties.

TABLE 1—*List of analyses evaluated during round-robin tests.*

pH	Iron
Conductivity	Lead
Alkalinity	Lithium
Hardness	Magnesium
Suspended solids	Manganese
Turbidity	Mercury
Aluminum	Phosphate
Ammonia	Potassium
Antimony	Rubidium
Arsenic	Silica (SiO$_2$)
Bicarbonate, carbonate	Silver
Barium	Sodium
Boron	Strontium
Bromide	Sulfate
Calcium	Sulfide
Cesium	Zinc
Chloride	Total dissolved solids
Copper	H$_2$S (total)
Fluoride	CO$_2$ (total)
Iodide	Radioactive isotopes

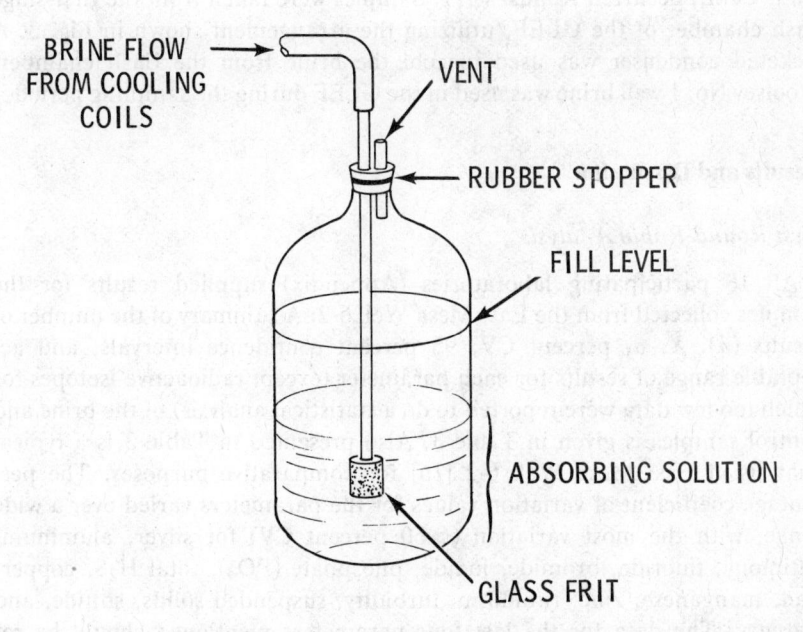

FIG. 1—*Carboy containing absorbing solution to trap gas (CO_2 or H_2S).*

gal) carboy. A similar arrangement was used for collecting hydrogen sulfide, with 0.5 M zinc acetate solution being used to absorb H_2S. The individual laboratory sample containers consisted of 1-litre linear polyethylene bottles, which had been acid washed and rinsed with distilled deionized water. Silica dilution bottles, consisting of 118-ml (4-oz) polypropylene bottles with Poly-Seal caps, had been previously filled with distilled deionized water. Brine was added to these to make 10/1 or 20/1 dilutions.

The makeup of each participant's sample set for the two round robins is presented in Table 2. The participants were given additional samples from the 49.2-litre (13-gal) carboys if it was requested for the purpose of completing the analysis.

Sample Collection

The first round-robin samples were collected in July 1977 at the East Mesa 6-2 wellhead site in the Imperial valley, Calif., using the sampling apparatus shown in Fig. 2. Brine from the well was cooled and condensed through two sets of cooling coils, as shown. Filtered samples were collected by passing the brine through a 0.45-μm filter and then into a 49.2-litre (13-gal) carboy.

Sampling for the second round-robin samples at the Geothermal Loop Experimental Facility (GLEF) of the San Diego Gas and Electric Co., Niland, Calif., occurred August 1977. Samples were taken from the first-stage flash chamber of the GLEF, utilizing the arrangement shown in Fig. 3. A jacketed condenser was used to cool the brine from the flash chamber. Woolsey No. 1 well brine was used in the GLEF during the sampling period.

Results and Discussion

First Round-Robin Analysis

All 18 participating laboratories (Appendix) supplied results for the samples collected from the East Mesa Well 6-2. A summary of the number of results (n), $\overline{X}$, σ, percent CV, 95 percent confidence intervals, and acceptable range of results for each parameter (except radioactive isotopes for which too few data were reported to do a statistical analysis) of the brine and control samples is given in Table 3. Also presented in Table 3 is a typical analysis of East Mesa Well 6-2 [16] for comparative purposes. The percentage coefficient of variation values for the parameters varied over a wide range, with the most variation (>50 percent CV) for silver, aluminum, antimony, fluoride, bromide, iodide, phosphate (PO_4), total H_2S, copper, lead, manganese, zinc, rubidium, turbidity, suspended solids, sulfide, and mercury. The data for the last four parameters mentioned should be regarded with caution because on-site measurement for turbidity, suspended solids, sulfide, and mercury would be preferable. No special precautions

TABLE 2—*Makeup of sample set for each participating laboratory.*

Sample Type	Round Robin No. 1	Round Robin No. 2
Raw unacidified (RU)	X	X
Raw acidified (RA)		X
Filtered unacidified (FU)	X	
Filtered acidified (FA)	X	
Dilution sample for SiO_2	X	X
Total CO_2 (collected in $2\,N$ NaOH solution)	X	X
Total H_2S (collected in $0.5\,M$ zinc acetate solution)	X	X
RA control		X
FA control	X	
SiO_2 control	X	X
CO_2 control	X	X
H_2S control	X	X
NaOH solution blank (solution used for absorbing CO_2)	X	X

were made to preserve a separate sample for mercury analysis, as has been proposed elsewhere [17].

The results confirm some previous findings [13] obtained with geothermal brines for parameters, such as bromide and iodide, where variability has been noted. The silica and sulfate (SO_4) determinations appeared to be better than those reported from the previous study [13] and appear to be reliable values, as they fall within the 95 percent confidence interval (CI). Fair agreement with control sample values was also achieved for both silica and sulfate, as can be seen from Table 3. The oxidation-titration, ion chromatographic, and chromic acid oxidation-extraction methods gave results for bromide within the 95 percent CI, as evidenced by the information in Table 8. Specific ion electrode measurements of bromide were of poor quality (Table 8). Procedures that gave results for iodide within the 95 percent CI included (from Table 8) oxidation-titration, specific ion electrode, arsenious ceric colorimetric, and arsenious-ceric–ferric-thiocyanate colorimetric methods. The poor analytical results for fluoride are due to values determined by procedures other than the specific ion electrode method, which gave values within the 95 percent CI (Table 8).

The poor results for the heavy metals silver, antimony, copper, lead, manganese, and zinc are probably partially attributable to their presence in trace concentrations in the brine. Reasonable agreement with control sample values was achieved for silver, antimony, copper, lead, manganese, and zinc, as can be seen from an examination of Table 3. This could be partly due to laboratory bias, as the approximate concentrations in the control samples were supplied by PNL. The aluminum results, broken down by method for the brine and control samples, are presented in Table 4. The various analytical methods that were used are presented along with their respective

FIG. 2—*GRR-1 sampling system.*

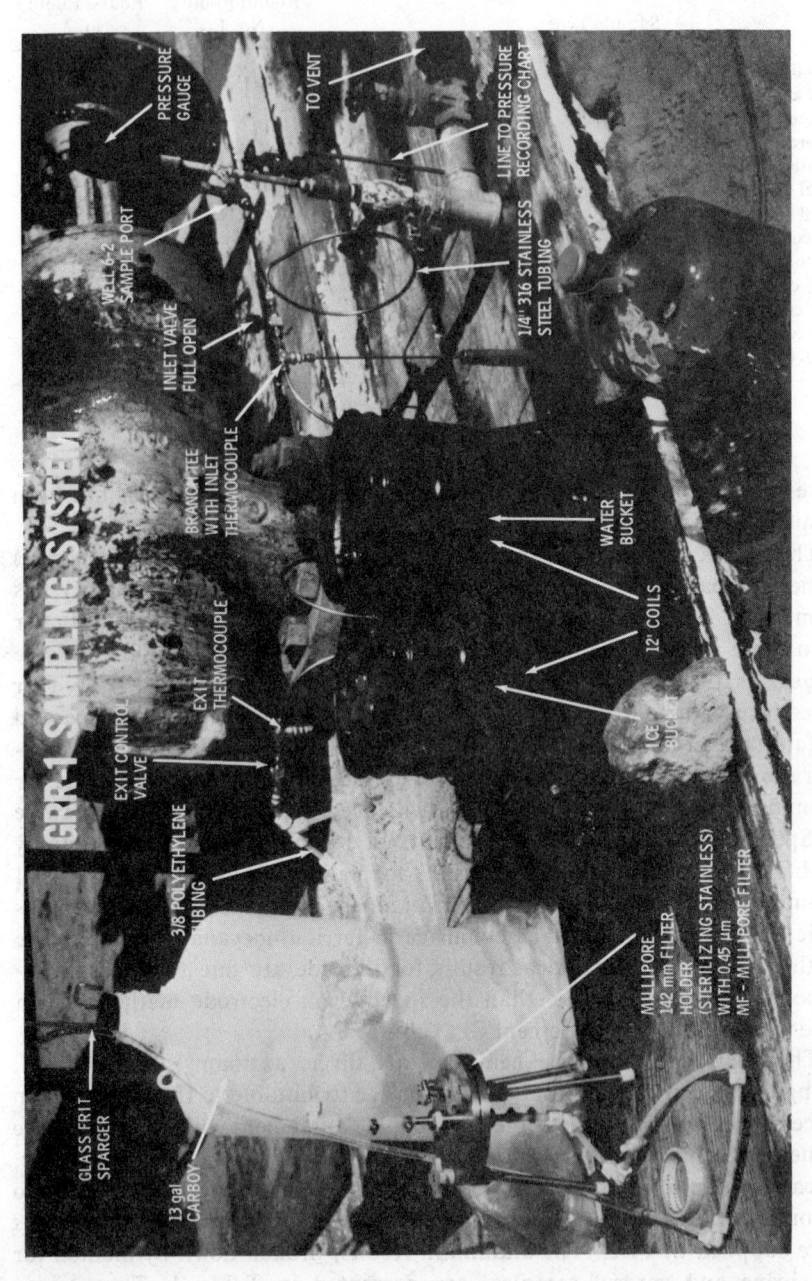

FIG. 3—*GRR-2 sampling system.*

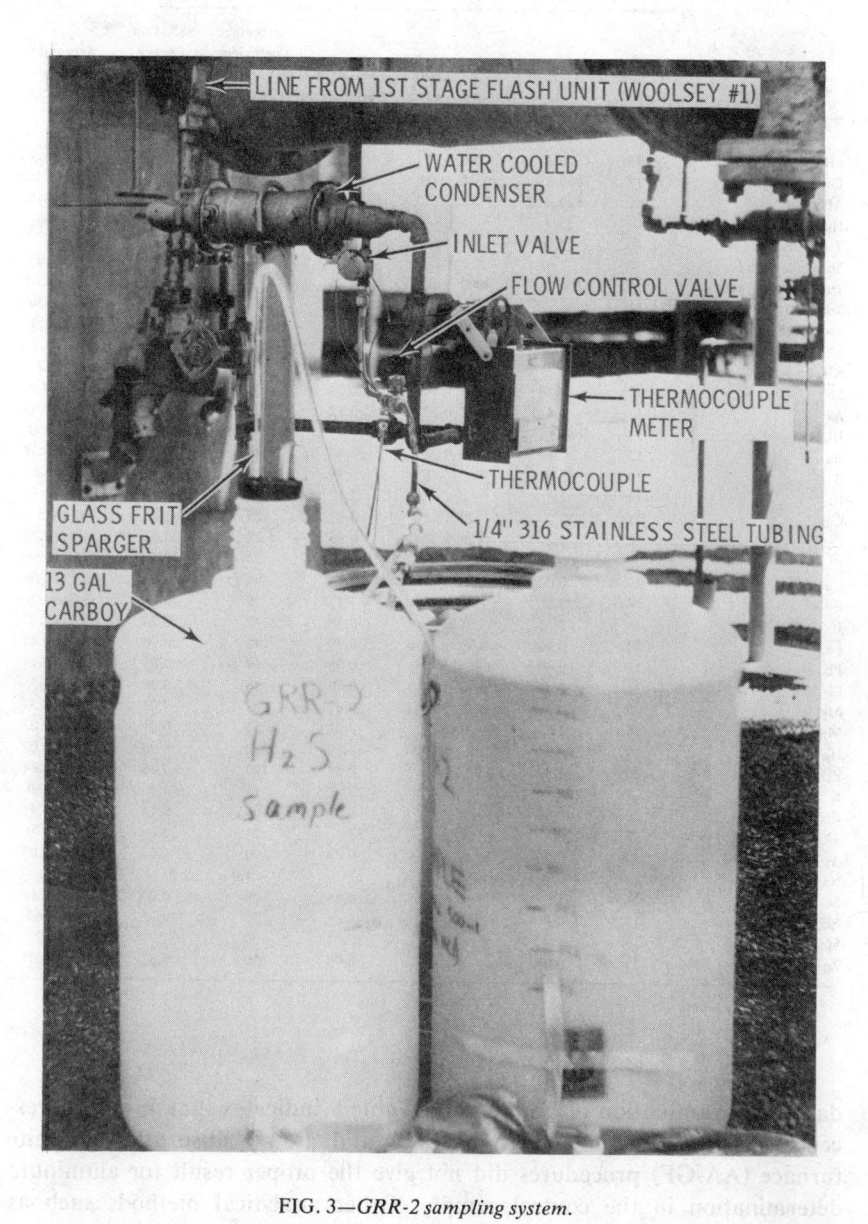

TABLE 3—*Results for Round Robin No. 1 sample and control.*

Brine Sample

Parameter	Number of Results, n	Average $\overline{X}$	Typical [16] Analysis	Standard Deviation, σ	Percentage Coefficient of Variation %CV	95% Confidence Interval about $\overline{X}$ ±	Acceptable Range of Values
pH	18	6.3	5.9, 6.12	0.36	6.0	0.17	6.1 to 6.5
Conductivity[a]	12	7000	6510, 6000	580	8.0	330	6700 to 7300
Alkalinity	10	520	...	36	7.0	22	500 to 540
Hardness	14	35	...	8.6	25	4.5	31 to 40
Turbidity[b]	8	2.9	...	6.1	220	4.2	0 to 7.1
Suspended solids	10	7.6	...	9.5	130	5.9	1.7 to 14
Total dissolved solids	14	4200	4000, 5000	150	3.0	79	4100 to 4300
Total CO_2	11	4600	...	1500	33	890	3700 to 5500
Total H_2S	11	2.0	1.6, 1.5	1.5	75	0.89	1.1 to 2.9
Al	17	2.3	0.03	5.8	250	2.8	0 to 5.1
NH, (as NH_4)	18	16	13, 14.7	2.1	13	0.97	15 to 17
Sb	15	0.18	0.02, 0.90	0.24	130	0.12	0.06 to 0.30
As	13	0.42	0.35, 0.22	0.12	28	0.07	0.35 to 0.49
HCO_3	13	600	744, 560	110	18	60	540 to 660
Ba	17	0.30	0.20, 0.25	0.10	34	0.05	0.25 to 0.35
B	19	8.0	9.65, 7.45	0.99	12	0.45	7.6 to 8.5
Br	14	2.6	...	2.1	79	1.1	1.5 to 3.7
Ca	21	13	20, 16.4	3.8	29	1.6	11 to 15
Cs	22	0.27	0.28, 0.38	0.08	28	0.03	0.24 to 0.30
Cl	23	1700	1920, 2142	560	33	230	1500 to 1900
Cu	15	0.04	<0.10	0.04	95	0.02	0.02 to 0.06
F	14	3.6	1.5, 1.23	1.1	30	0.58	3.0 to 4.2
I	8	0.53	...	0.35	67	0.24	0.29 to 0.77
Fe	24	0.40	0.90, <0.10	0.19	48	0.08	0.32 to 0.48
Pb	11	0.09	0.004, <0.5	0.08	90	0.05	0.04 to 0.14
Li	21	4.4	0.40, 4.0	0.70	16	0.30	4.1 to 4.7
Mg	20	0.31	0.47, 0.24	0.07	23	0.03	0.28 to 0.34
Mn	16	0.02	0.05	0.01	53	0.01	0.01 to 0.03
Hg	10	0.03	0.0038, <0.002	0.09	310	0.06	0 to 0.09
PO_4	15	0.39	0.05, <0.2	0.49	120	0.25	0.14 to 0.64
K	27	140	155, 150	25	18	9.4	130 to 150
Rb	20	0.98	...	0.97	99	0.43	0.55 to 1.4
SiO_2	25	250	220, 269	55	22	22	230 to 270
Ag	10	0.03	<0.01	0.03	99	0.02	0.01 to 0.05
Na	28	1400	1450, 1700	61	4.0	23	1400
Sr	24	1.8	1, 6.4	0.68	38	0.27	1.5 to 2.1
SO_4	13	140	160, 156	45	31	24	120 to 160
Sulfide	13	25	...	50	200	27	0 to 52
Zn	23	0.05	<0.01	0.06	120	0.02	0.03 to 0.07

[a] Units = μmhos/cm.

data. An examination of Table 4 and Table 8 indicates that X-ray fluorescence (XRF), neutron activation (NA), and atomic absorption–graphite furnace (AA-GF) procedures did not give the proper result for aluminum determination in the control sample. Other analytical methods such as atomic absorption (AA), flame emission (FE), inductively coupled plasma atomic-emission spectroscopy (ICP-AES), spark-source mass spectroscopy (SSMS), and others performed much better.

East Mesa 6-2 wellhead brine (concentrations in mg/litre).

Number of Results, n	Average $\overline{X}$	PNL Control Value (as prepared)	Standard Deviation, σ	Percentage Coefficient of Variation %CV	95% Confidence Interval about $\overline{X}$, ±	Acceptable Range of Values
			Control Sample			
11	34	42	9.3	27	5.5	29 to 40
9	2100	2000	210	10	140	2000 to 2200
10	5.5	5.5	2.3	42	1.4	4.1 to 6.9
16	1.1	0.80	0.47	43	0.23	0.87 to 1.3
12	13	14	3.2	25	1.8	11 to 15
13	0.84	1.0	0.30	36	0.16	0.68 to 1.0
8	0.28	0.30	0.10	34	0.07	0.21 to 0.35
17	0.32	0.30	0.07	21	0.03	0.29 to 0.35
11	7.3	7.0	2.0	27	1.2	6.1 to 8.5
9	11	5.0	16	150	10	1.0 to 21
18	13	13	1.9	15	0.88	12 to 14
13	0.43	0.42	0.11	24	0.06	0.37 to 0.49
14	0.09	0.08	0.03	33	0.02	0.07 to 0.11
8	0.61	1.2	0.38	63	0.26	0.35 to 0.87
7	4.1	5.0	1.6	38	1.2	2.9 to 5.3
17	0.76	0.70	0.16	21	0.08	0.68 to 0.84
14	0.35	0.30	0.14	40	0.07	0.28 to 0.42
15	4.0	4.0	0.78	19	0.40	3.6 to 4.4
16	0.20	0.20	0.06	29	0.03	0.17 to 0.23
13	0.06	0.06	0.01	12	0.01	0.05 to 0.07
4	0.01	0.01	0	71	...	0.01
13	0.29	0.25	0.16	54	0.09	0.20 to 0.38
18	180	180	26	15	12	170 to 190
13	5.3	5.0	1.4	26	0.76	4.5 to 6.1
16	210	250	89	41	44	170 to 250
11	0.04	0.07	0.04	98	0.02	0.02 to 0.06
19	1500	1500	60	4.0	27	1500
16	1.5	1.6	0.16	11	0.08	1.4 to 1.6
6	160	150	17	11	14	150 to 170
17	0.63	0.60	0.08	13	0.04	0.59 to 0.67

[b]Units = turbidity units.

Results for phosphate in the East Mesa Well 6-2 brine showed this parameter to be present as a minor constituent. An examination of Table 8 for the brine and control samples indicates that the colorimetric procedures using ascorbic acid or stannous chloride reduction, and ICP-AES gave the most results within the 95 percent CI.

Analysis for rubidium by flame emission, atomic absorption, or neutron activation analysis gave results within the 95 percent CI.

TABLE 4—*Statistical evaluation of aluminum results for Round Robin No. 1.*[a]

Pt	Method	Result	Lower −5σ	−3σ	Xbar	Upper 3σ	5σ
		Aluminum unknown, (mg/litre[b])					
1	AA	0.24			*		
2	AA	0.11			*		
3	AA	0.0			*		
4	AA EXT	0.16			*		
5	FE	0.20 RA			*		
6	FE	0.20 FA			*		
7	AA GF	0.22			*		
8	AA GF	0.13			*		
9	ICP-AES	0.16			*		
10	ICP-AES	0.30			*		
11	ICP-AES	0.07			*		
12	ES	0.57			*		
13	SSMS	0.20 FA			*		
14	SSMS	0.06			*		
15	XRF	37.00 FA					*
16	XRF	15.00 FU				*	
17	XRF	19.00 RU				*	
		Aluminum Control 0.80 mg/litre[c]					
1	AA	1.00			*		
2	AA	1.00			*		
3	AA	1.00			*		
4	AA	1.70			*		
5	AA EXT	0.94			*		
6	FE	1.00			*		
7	AA GF	0.81			*		
8	AA GF	0.52			*		
9	ICP-AES	0.96			*		
10	ICP-AES	1.30			*		
11	ICP-AES	0.59			*		
12	ES	2.00				*	
13	ES	0.87			*		
14	NA	2.00				*	
15	SSMS	0.60			*		
16	XRF	43.00					*

[a] Abbreviations used:
AA = Atomic absorption.
AA EXT = Extraction followed by AA.
FE = Flame emission.
AA GF = Atomic absorption, graphite furnace.
ICP-AES = Inductively coupled plasma-atomic emission spectroscopy.
ES = Emission spectroscopy.
NA = Neutron activation analysis.
SSMS = Spark source mass spec.
XRF = X-ray fluorescence.
RA = Raw acidified.
RU = Raw unacidified.
FA = Filtered acidified.
FU = Filtered unacidified.

[b] Chart of individuals, xbar and 3σ limits based on Points 1 to 17 omitting outliers. $N = 16$; $\sigma = 5.790$; %RSO = 253; lower = 15.082; upper = 19.660; xbar = 2.289.

[c] Based on Points 1 to 16 omitting outliers. $N = 15$; $\sigma = 0.469$; %RSO = 43; lower = −0.322; upper = 2.494; xbar = 1.086.

The total hydrogen sulfide analysis gave equivalent results for both the iodometric titration and the methylene-blue colorimetric methods [14].

Second Round-Robin Analysis

Seventeen laboratories reported results for the second round-robin set of samples. The agreement of results was, as expected, less than that for the first round-robin analysis. The higher TDS (24 percent) of the brine, as well as the associated matrix effects and chemical interferences, complicated the analysis. Unlike in the first round-robin, the participating laboratories had no advance information concerning the makeup of the control sample.

The results, as reported by the laboratories, were tabulated and n, $\overline{X}$, σ, percent CV, 95 percent confidence intervals, and the acceptable range of results were calculated for each parameter (Table 5). As was the case with the first round robin, not enough results were available to perform statistical evaluation of the radioactive isotope data. The percentage coefficient of variation values were higher (>50 percent CV) for many of the same parameters as was true in the first round robin, such as silver, aluminum, antimony, fluoride, bromide, iodide, PO_4, and total H_2S. Table 8 presents the methods that gave the most reliable results for each parameter. Many methods that gave results within the 95 percent CI for the first round-robin samples did not do as well on the higher-TDS brine, and, conversely, many methods that failed on the first round-robin samples performed adequately on the high-TDS brine samples. The parameters of turbidity, suspended solids, sulfide, and mercury also had percent CV values greater than 50. As with the first round robin, these analyses should be determined on site. The large amount of ferrous iron present in the Woolsey No. 1 brine can account for some of the anomalous results noted for pH (Table 5) and suspended solids. The ferrous iron slowly converts to ferric ion on standing, which leads to copious precipitation of ferric oxide and a corresponding reduction in pH. Field pH measurements taken at the time of collection were ~6.2. Also of note is the precipitation of barium sulfate and silica in the sample due to the cooling of the saturated solution. An examination of Table 5, which contains a reported analysis [17] of the first-stage flash fluid, shows a decrease in the concentration of barium. In addition, the zinc average value, $\overline{X}$, is significantly different from the reported value, probably as a result of zinc precipitation. The agreement with the PNL control value indicates that the problem was not due to laboratory measurement for zinc.

Additional parameters exhibiting high percent CV values were barium, bicarbonate, (HCO_3), boron, potassium, SO_4, arsenic, and total CO_2. As mentioned previously, all the barium and SO_4 concentrations reported are questionable because of the preceding argument. The bicarbonate concentration variation could conceivably be due to the high borate and silicate concentrations, which give apparent bicarbonate concentrations, as has been previously reported [13]. Boron determinations (Table 8), using the cur-

TABLE 5—*Results for Round Robin No. 2 sample and control.*

Brine Sample

Parameter	Number of Results, n	Average, $\overline{X}$	Typical Analysis [17]	Standard Deviation, σ	Percentage Coefficient of Variation %CV	95% Confidence Interval about $\overline{X} \pm$	Acceptable Range of Values
pH	13	3.7	...	0.70	19	0.38	3.3 to 4.1
Conductivity[a]	7	220 00	...	120 000	53	90 000	130 000 to 310 000
Hardness	8	60 000	...	2 000	3.0	1 400	59 000 to 61 000
Turbidity[b]	4	160	...	160	100	160	0 to 320
Suspended Solids	5	1 400	...	930	68	810	590 to 2 200
Total dissolved solids	10	240 000	...	18 000	8.0	11 000	230 000 to 250 000
Total CO_2	9	510	...	600	120	390	120 to 900
Total H_2S	4	0.94	...	0.88	94	0.86	0.08 to 1.8
Al	13	2.4	...	2.4	99	1.3	1.1 to 3.7
NH_3 (as NH_4)	14	360	...	28	8.0	15	350 to 380
Sb	13	2.3	...	3.6	160	2.0	0.30 to 4.3
As	16	8.0	...	5.8	72	2.8	5.2 to 11
HCO_3	8	0.63	...	1.8	280	1.2	0 to 1.8
Ba	18	120	145	63	53	29	91 to 150
B	15	460	...	450	98	230	230 to 700
Br	10	260	...	380	150	240	20 to 500
Ca	21	23 000	21 300	4 900	21	2 100	21 000 to 25 000
Cs	18	9.4	...	4.3	46	2.0	7.4 to 11
Cl	15	130 000	121 200	5100	4.0	2 600	130 000
Cu	18	1.1	0.8	0.44	40	0.20	0.90 to 1.3
F	12	7.9	...	12	160	6.8	1.1 to 15
I	10	8.7	...	7.5	86	4.7	4 to 13
Fe	24	260	225	46	17	18	240 to 280
Pb	18	42	43	19	45	8.8	33 to 51
Li	19	170	...	19	11	8.5	160 to 180
Mg	20	150	160	20	13	8.8	140 to 160
Mn	16	690	600	96	14	47	640 to 740
Hg	7	0.01	...	0.01	230	0.01	0 to 0.02
PO_4	8	1.9	...	2.5	130	1.7	0.2 to 3.6
K	21	14 000	10 500	12 000	90	5 100	8 900 to 19 000
Rb	18	65	...	22	33	10	55 to 75
SiO_2	22	360	364	82	23	34	330 to 390
Ag	10	0.53	...	0.75	140	0.47	0.06 to 1.0
Na	24	47 000	43 600	4 500	9.0	1 800	45 000 to 49 000
Sr	19	360	...	130	38	58	300 to 420
SO_4	12	66	...	54	81	31	35 to 97
Sulfide	5	7.5	...	14	190	12	0 to 20
Zn	21	220	135	94	42	40	180 to 260

[a]Units = μmhos/cm.

cumin colorimetric, FE, carmine colorimetric, SSMS, ICP-AES, and emission spectroscopy (ES) methods gave results within the 95 percent CI.

The variability in potassium results is obvious from an examination of Table 6, which shows the various analytical procedures used for potassium. The flame emission and absorption values agree reasonably well for the brine sample, with the exception of two flame emission values at 3σ from the average value. For the control sample, the data scatter much more for both flame emission and absorption values, with one emission value at 5σ from the average value.

The analytical results for arsenic are much worse for the control sample

Woolsey No. 1 flashed brine, Concentrations in mg/litre.

Number of Results, n	Average, X	PNL Control Value (as prepared)	Control Sample Standard Deviation, σ	Percentage Coefficient of Variation %CV	95% Confidence Interval about $\overline{X}$, ±	Acceptable Range of Values
9	1 400	600	1 400	100	920	480 to 2300
10	2.2	2.0	1.2	57	0.74	1.5 to 2.9
12	2.2	0.98	1.6	74	0.91	1.3 to 3.1
11	410	450	98	24	58	350 to 470
10	4.1	4.3	3.9	95	2.4	1.7 to 6.5
8	1.4	0.10	3.1	210	2.1	0 to 4.5
14	78	90	18	23	9.4	69 to 87
12	180	210	68	37	39	140 to 220
5	270	1.6	350	130	310	0 to 580
19	13 000	13 000	2 000	16	900	12 000 to 14 000
14	150	160	68	46	36	110 to 190
16	2.0	1.9	0.70	35	0.34	1.7 to 2.3
8	17	12	34	200	24	0 to 41
7	4.6	3.3	3.8	84	2.8	1.8 to 7.4
19	120	100	24	20	11	110 to 130
16	21	23	7.1	34	3.5	18 to 25
17	96	98	29	30	14	82 to 110
19	110	110	20	18	9.0	100 to 120
15	350	340	65	19	33	320 to 380
7	1.1	1.1	0.29	27	0.22	0.88 to 1.3
9	0.53	0.46	0.38	71	0.25	0.28 to 0.78
19	6 200	6 900	1 200	19	540	5 700 to 6 700
13	38	39	12	31	6.5	32 to 45
18	36	45	8.3	23	3.8	32 to 40
11	0.45	0.13	0.47	100	0.28	0.17 to 0.73
19	17 000	18 000	1 600	9.0	720	16 000 to 18 000
16	150	150	50	35	25	130 to 180
7	18	0.00	31	190	23	0 to 41
17	89	90	35	39	17	72 to 110

[b]Units = turbidity units.

than for the brine, although the scatter is more pronounced in the brine sample (Table 7). Of the methods used (Table 8), hydride evolution (silver diethyldithiocarbamate colorimetric procedure), AA-graphite furnace analysis, ICP-AES, spark source mass spectroscopy, and atomic absorption determination of arsine gave reliable results.

The total carbon dioxide analysis values appear to have been calculated with a dilution factor in error, as the results obtained for the control sample are roughly twice the value of the PNL-prepared control sample (as prepared). This error, which probably arose during computation, points out the necessity of good data-handling procedures.

TABLE 6—*Statistical evaluation of potassium results for Round Robin No. 2.*[a]

Pt	Method	Result	Lower −5σ	−3σ	Xbar	Upper 3σ	5σ
		Potassium Unknown, mg/litre[b]					
1	AA	10 500			*		
2	AA	10 000			*		
3	AA	6 850			*		
4	AA	10 270			*		
5	AA	12 300			*		
6	AA	10 200			*		
7	AA	11 000			*		
8	FE	11 300			*		
9	FE	8 000			*		
10	FE	10 000			*		
11	FE	10 800			*		
12	FE	50 000 RA				*	
13	FE	50 000 RU				*	
14	FE	10 400			*		
15	FE	10 300			*		
16	ICP-AES	8 500			*		
17	ES	4 200			*		
18	ES	9 100			*		
19	ION EXC	11 300 RU			*		
20	ION EXC	11 200 RA			*		
21	NA	9 025 RU			*		
		Potassium Control 6910, mg/litre[c]					
1	AA	6 900			*		
2	AA	6 750			*		
3	AA	3 000		*			
4	AA	6 650			*		
5	AA	6 920			*		
6	AA	6 700			*		
7	AA	7 200			*		
8	FE	6 200			*		
9	FE	5 000			*		
10	FE	6 620			*		
11	FE	6 450			*		
12	FE	25 000					*
13	FE	7 000			*		
14	FE	5 150			*		
15	ICP-AES	6 400			*		
16	ES	4 200		*			
17	ES	6 600			*		
18	ION EXC	7 150			*		
19	NA	7 500			*		

[a] Abbreviations used:
  AA = Atomic absorption.
  FE = Flame emission.
  ICP-AES = Inductively coupled plasma-atomic emission spectroscopy.
  ES = Emission spectroscopy.
  ION EXC = Ion chromatography.
  NA = Neutron activation analysis.
  RU = Raw unacidified.
  RA = Raw acidified.

[b] Chart of individuals Xbar and 3σ limits based on Points 1 to 21 omitting outliers. $N = 21$; $\sigma = 12235.9$; %RSD = 90; lower = 23 124.5; upper = 50 290.7; Xbar = 13 583.1.

[c] Based on Points 1 to 19 omitting outliers. $N = 18$; $\sigma = 1167.8$; %RSD = 19; lower = 2740.6; upper = 9747.2; Xbar = 6243.9.

TABLE 7—*Statistical evaluation of arsenic results for Round Robin No. 2.*[a]

Pt	Method	Result	Lower −5σ	−3σ	Xbar	Upper 3σ	5σ
		Arsenic Unknown, mg/litre[b]					
1	AA GF	0.04			*		
2	AA GF	10.40			*		
3	AA GF	0.18			*		
4	H EV SP	6.40			*		
5	H EV SP	10.00			*		
6	ICP-AES	5.90			*		
7	ICP-AES	22.00				*	
8	ICP-AES	11.00			*		
9	ES	3.50			*		
10	NA	10.50 RA			*		
11	NA	4.60 RU			*		
12	SSMS	7.00 RA			*		
13	SSMS	7.00 RU			*		
14	SSMS	17.00				*	
15	AA H EV	10.50			*		
16	AA H EV	2.60			*		
		Arsenic Control 0.098, mg/litre[c]					
1	AA GF	0.02			*		
2	AA GF	0.80			*		
3	H EV SP	0.30			*		
4	ICP-AES	9.00				*	
5	ICP-AES	0.23			*		
6	SSMS	1.00			*		
7	AA H EV	0.10			*		
8	AA H EV	0.03			*		

[a]Abbreviations used:
AA GF = Atomic absorption, graphite furnace.
H EV SP = Hydride evolution and spectrophotometric (silver diethyldithiocarbamate).
ICP-AES = Inductively coupled plasma-atomic emission spectroscopy.
ES = Emission spectroscopy.
NA = Neutron activation analysis.
SSMS = Spark source mass spec.
AA H EV = Arsine and atomic absorption.
RA = Raw acidified.
RU = Raw unacidified.

[b]Chart of individuals, Xbar and 3σ limits based on Points 1 to 16 omitting outliers. $N = 16$; $\sigma = 5.789$; %RSD = 72; lower = −9.328; upper = 25.406; Xbar = 8.039.

[c]Based on Points 1 to 8 omitting outliers. $N = 8$; $\sigma = 3.078$; %RSD = 214; lower = −7.799; upper = 10.669; Xbar = 1.435.

Conclusions

For the low-TDS (0.4 percent) brine sample, 17 species showed coefficients of variation greater than 50 percent. Of these 17 species, 5 (copper, lead, manganese, rubidium, and zinc) were present in trace concentrations, which might explain the poor quality of the data associated with them. Four other

252 GEOTHERMAL SCALING AND CORROSION

TABLE 8—*Summary of analysis methods used.*

Parameter	GRR-1 (Low TDS Brine and Control) Within 95% Confidence Interval	GRR-1 (Low TDS Brine and Control) Outside 95% Confidence Interval	GRR-2 (High TDS Brine and Control) Within 95% Confidence Interval	GRR-2 (High TDS Brine and Control) Outside 95% Confidence Interval
pH	pH meter	...	pH meter	...
Conductivity	meter	...	meter	...
Alkalinity	acid titration	...		
Hardness	EDTA titration calculation	...	EDTA titration, calculation	...
Turbidity	nephelometer, visual matching, Hach meter	...	nephelometer, Hach meter	...
Suspended solids	gravimetric	...	gravimetric	
TDS	gravimetric	...	gravimetric	
Total CO_2	acid titration, precise evolution, strontium gravimetric, carbon analyzer, calculation	...	acid titration, precise evolution	strontium gravimetric, calculation
Total H_2S	iodometric titration, methylene blue colorimetric	Lauth's violet colorimetric	methylene blue colorimetric	iodometric titration, Lauth's violet colorimetric
Al	AA, AA-extraction, FE, ICP-AES, ES, SSMS	XRF, NA, AA-graphite furnace	ES, SSMS	AA, FE, AA-graphite furnace, ICP-AES
NH_3	Auto Analyzer (colorimetric phenate), ion chromatography	nesslerization, specific ion electrode, acidimetric	Auto Analyzer (colorimetric phenate), ion chromatography	nesslerization, specific ion electrode, acidimetric
Sb	AA, NA	AA-hydride evolution, ES, AA-graphite furnace, XRF, ICP-AES, NA, SSMS	AA, ES, ICP-AES, NA, SSMS	AA-hydride evolution, AA-graphite furnace, XRF
As	hydride evolution, spectrophotometric, ES, arsine-AA, ES	AA-graphite furnace, ICP-AES, NA, SSMS	AA-graphite furnace, hydride evolution-spectrophotometric, ICP-AES, SSMS, arsine-AA	ES, NA

HCO$_3$	acid titration	calculation	acid titration, calculation	
Ba	AA, AA-graphite furnace, ES, FE, SSMS	ICP-AES, NA, absorptometric	AA, ICP-AES, SSMS	AA-graphite furnace, ES, FE, NA, absorptometric
B	curcumin colorimetric, FE, ICP-AES	carmine colorimetric, SSMS, ES, AA	curcumin colorimetric, FE, carmine colorimetric, SSMS, ICP-AES, ES	
Br	American Petroleum Institute photometric, oxidation-titration, ion chromatography	specific ion electrode, I$_2$-permanganate, XRF, NA, SSMS, phenol red colorimetric	API photometric, oxidation-titration, XRF, NA, ion chromatography, SSMS, specific ion electrode	I$_2$-permanganate
Ca	AA, ICP-AES, SSMS	NA, EDTA titration, ES, FE	AA, SSMS, FE, EDTA titration, ICP-AES	NA, ES
Cs	NA	AA, FE, SSMS, ES, AA-graphite furnace	NA	AA, SSMS, ES, AA-graphite furnace, FE
Cl	NA	AgNO$_3$ titration, Hg(NO$_3$)$_2$ titration, XRF, NA, gravimetric specific ion electrode, ion chromatography	AgNO$_3$ titration, Hg(NO$_3$)$_2$ titration, gravimetric, ion chromatography	XRF, specific ion electrode
Cu	ES	AA, AA-extraction, NA, SSMS, ICP-AES, AA-graphite furnace, XRF	AA, AA-graphite furnace ICP-AES	SSMS, ES
F	specific ion electrode	alizarin visual, AMADAC-F colorimetric, SSMS, ion chromatography, 4,5-dihydroxy-3[(p-sulfophenyl) azo]-2,7-naphthalenedisulfonic acid, trisodium salt colorimetric	specific ion electrode, alizarin visual, AMADAC-F colorimetric, ion chromatography. 4,5-dihydroxy-3[(p-sulfophenyl) azo]- 2,7-)naphthalenedisulfonic acid, trisodium salt colorimetric	SSMS

TABLE 8—Continued

Parameter	GRR-1 (Low TDS Brine and Control) Within 95% Confidence Interval	GRR-1 (Low TDS Brine and Control) Outside 95% Confidence Interval	GRR-2 (High TDS Brine and Control) Within 95% Confidence Interval	GRR-2 (High TDS Brine and Control) Outside 95% Confidence Interval
I	oxidation titration, specific ion electrode, arsenious ceric colorimetric, arsenious ceric Fe-thiocyanate colorimetric	NA, XRF, Leuco crystal violet colorimetric, SSMS, API photometric	oxidation titration, specific ion electrode, API photometric	XRF, leuco crystal violet colorimetric, SSMS, arsenious ceric colorimetric, arsenious ceric Fe-thiocyanate colorimetric
Fe	ES, AA	AA-extraction, ICP-AES, XRF, NA, SSMS	AA, ES, phenanthroline colorimetric	AA-extraction, SSMS, ICP-AES, NA, FE
Pb	AA, AA-graphite furnace, ICP-AES	SSMS, AA-extraction, ES	AA, ICP-AES	AA-graphite furnace, SSMS, ES
Li	AA, FE, ES, ion chromatography	SSMS	AA, FE, ES, ion chromatography	SSMS
Mg	AA, ICP-AES	AA, NA, SSMS, ES, EDTA titration	AA, ES, EDTA titration	SSMS, ICP-AES
Mn	AA, AA-extraction, SSMS, ICP-AES, ES	XRF, NA	AA, ES	SSMS, ICP-AES
Hg	cold vapor AA, XRF ICP-AES, colorimetric (SnCl$_2$ reduction), colorimetric (ascorbic acid reduction)	NA	cold vapor AA colorimetric (ascorbic acid reduction), molybdophosphoric acid colorimetric, SSMS	NA, ES colorimetric (SnCl$_2$ reduction)
PO$_4$		SSMS, XRF, molybdophosphoric acid colorimetric		
K	AA, FE, ion chromatography	SSMS, ICP-AES, ES, XRF, NA, specific ion electrode	AA, FE, ES, ion chromatography, NA	SSMS, ICP-AES
Rb	AA, FE, NA	SSMS, ES, AA-graphite furnace	AA, FE, NA, ES	AA-graphite furnace, SSMS
SiO$_2$	heteropoly blue colorimetric, ICP-AES	AA, molybdosilicate colorimetric, FE, SSMS, ES	molybdosilicate colorimetric	AA, heteropoly blue colorimetric, FE, ICP-AES, SSMS, gravimetric

Ag	NA, ICP-AES	AA, AA-graphite furnace AA-extraction, SSMS	ES, NA, SSMS
Sr	AA, FE, SSMS, ICP-AES NA, AA-graphite furnace	ES, XRF	SSMS, ES, AA-graphite furnace
SO$_4$	gravimetric-ignition, turbidimetric, gravimetric-drying, ion chromatography	AA, FE, ICP-AES, NA gravimetric-ignition, turbidimetric gravitimetric-drying	AA, FE, ion chromatography
Sulfide	iodometric titration, methylene blue colorimetric, XRF, specific ion electrode	NA, SSMS iodometric titration, methylene blue colorimetric	SSMS, potassium-Sb-tartrate qualitative
Zn	AA, ICP-AES	AA-extraction ES, SSMS, AA-graphite furnace, NA	ES, SSMS, ICP-AES AA, AA-extraction, AA-graphite furnace, NA

analyses (for turbidity, suspended solids, sulfide, and mercury) should be determined on site, or, in the case of mercury, the sample should be preserved differently. Results for bromide and iodide confirm previous findings where variability has been noted. The scatter of fluoride results is attributable to the use of methods other than the specific ion electrode, which gave results within the 95 percent CI. Silica and sulfate were easier to determine than had been reported in previous studies.

For the high-TDS (24 percent) brine, 19 species had percent CV values greater than 50. The components silver, aluminum, antimony, fluoride, bromide, iodide, PO_4, and total H_2S, fell into this category, as did the first round-robin samples. In addition, the four analyses—turbidity, suspended solids, sulfide, and mercury—common to both round-robin sets had percent CVs >50 and should be determined on site, or, in the case of mercury, the sample should be preserved differently. Barium and SO_4 showed variability in results due to precipitation. Variability in HCO_3 values could be attributable to high silicate and borate concentrations. The scatter for boron, arsenic, and potassium values is disturbing because the first two values are important environmentally, and reliable potassium data are needed to calculate reservoir temperatures based on sodium/potassium ratios [19]. The error in CO_2 values is thought to have arisen from improper calculations, as the scatter for CO_2 was not observed during the first round robin.

A table of analysis procedures which gave results within the 95 percent CI for each parameter is presented (Table 8). Examination shows that many procedures found to be inadequate for the low-TDS brine-control samples, such as SSMS, ES, and others, gave reliable results for the high-TDS brine-control samples. Conversely, methods such as AA, ICP-AES, and others that performed poorly on the high-TDS brine-control samples gave reliable results for the low-TDS brine-control samples.

Acknowledgments

I wish to thank Mark A. Floyd of the Ames Laboratory, Energy and Mineral Resources Research Institute, U.S. Department of Energy, Ames, Iowa, for performing the statistical evaluation of the data from the round-robin samples. I would like to acknowledge the U.S. Department of Energy, Division of Geothermal Energy, Utilization Branch, Washington, D.C., for providing funding for this study. Acknowledgment is also extended to the organizations that participated in the round-robin study and that provided the data on which this paper is based.

The effort to standardize methods of analysis through ASTM is continuing at PNL. This work is funded by the U.S. Department of Energy, Division of Geothermal Energy. Donald W. Shannon at PNL is coordinating this effort.

APPENDIX

Brine Analysis Round-Robin Participants

Government Agencies	Government-Funded Laboratories	Private Laboratories	Universities	Industries
Environmental Protection Agency,[a,b,c] Environmental Monitoring and Support Laboratory, Las Vegas, Nev.	Ames Laboratory,[a,b,c] Ames, Iowa	Atomics International,[a,b,c] Canoga Park, Calif.	University of Southern California,[b,c] Los Angeles, Calif.	Allied Chemical Corp.,[a,b,c] Idaho Falls, Idaho
Oak Ridge National Laboratory,[a,b,c] Oak Ridge, Tenn.	Los Alamos Scientific Laboratory,[a] Los Alamos, N. Mex.	LFE Laboratories[a,b,c] Richmond, Calif.		Chevron Oil Co.,[a,b,c] LaHabra, Calif.
U.S. Bureau of Mines,[b,c] College Park Metallurgy Research Center, College Park, Md.	Lawrence Livermore Laboratory,[a] Livermore, Calif.	Physical Dynamics,[a,b,c] Mercer Island, Wash.		Union Oil Co.,[a,b,c] Brea, Calif.
U.S. Geological Survey,[b,c] Menlo Park, Calif.	Pacific Northwest Laboratory,[a,b,c] Richland, Wash.	TRW,[a,b,c] Redondo Beach, Calif.		Westec Services[c] Niland, Calif.
	Gulf South Research Institute,[b,c] New Orleans, La.	Vetter Research,[a,c] Costa Mesa, Calif.		
	Lawrence Berkeley Laboratory,[b,c] East Mesa, Calif.	GHT Laboratory,[b,c] Brawley, Calif.		
	Radian Corp.,[c] Austin, Tex.			

[a] Organizing workshop, May 1977.
[b] Geothermal round robin 1 (GRR-1) participants.
[c] Geothermal round robin 2 (GRR-2) participants.

References

[1] Douglas, J. G., Serne, R. J., Shannon, D. W., and Woodruff, E. M., "Geothermal Water and Gas—Collected Methods for Sampling and Analysis, Comment Issue," Technical Report BNWL-2094, Battelle, Pacific Northwest Laboratories, Richland, Wash., 1976.
[2] Ekedahl, G. and Rondell, B., *Vatten,* Vol. 29, 1973, pp. 341-356.
[3] Rondell, B., *Vatten,* Vol. 29, 1973, pp. 357-365.
[4] Henriksen, A., *Vatten,* Vol. 31, 1975, pp. 91-93.
[5] Brown, E., Skougstad, M. W., and Fishman, M. J., *Methods for Collection and Analysis of Water Samples for Dissolved Minerals and Gases, Techniques of Water Resources Investigations,* Book 5, Chapter A1, U.S. Geological Survey, Stock No. 2401-1015, U.S. Printing Office, Washington, D.C., 1970.
[6] Lishka, R. J. and McFarren, E. F., "Water Physics No. 1," Report No. 39, U.S. Environmental Protection Agency, Washington, D.C. 1971.
[7] Kingsford, M., Stevenson, C. D., and Edgerley, W. H. L., *New Zealand Journal of Science,* Vol. 16, No. 4, 1973, pp. 895-902.
[8] Stevenson, C. D., Kingsford, M., and Edgerley, W. H. L., *New Zealand Journal of Science,* Vol. 19, No. 4, 1976, pp. 353-357.
[9] Kingsford, M., Stevenson, C. G., and Edgerley, W. H. L., "Collaborative Tests of Water Analysis (the CHEMAQUA Program): III. Sodium, Potassium, Calcium, Magnesium, Chloride, Sulfate, Bicarbonate, Silica, Conductivity, and pH, *New Zealand Department of Scientific and Industrial Research,* Chemistry Div., Chemistry Div. Report No. 2242, New Zealand, 1977.
[10] Smith, R., *Water SA,* Vol. 3, No. 2, 1977, pp. 66-71.
[11] Dybczynski, R., Tugsavul, A., and Suschny, O., *Analyst,* Vol. 103, No. 1228, 1978, pp. 734-744.
[12] Sugawara, K., *Deep-Sea Research,* Vol. 25, No. 3, 1978, pp. 323-332.
[13] Ellis, A. J., *Geochimica et Cosmochimica Acta,* Vol. 40, 1976, pp. 1359-1374.
[14] Franson, M. A., *Standard Methods for the Examination of Water and Wastewater,* American Public Health Association, Washington, D.C., 1976.
[15] Watson, J. C., "Sampling and Analysis Methods for Geothermal Fluids and Gases," Technical Report PNL-MA-572, Battelle, Pacific Northwest Laboratory, Richland, Wash., 1979; now available through the National Technical Information Service (NTIS), Springfield, Va.
[16] "Geothermal Resource Investigations, East Mesa Test Site, Imperial Valley, California," Status Report, U.S. Department of the Interior, Bureau of Reclamation, Boulder City, Nev., April 1977.
[17] Bishop, H. K., Bricarello, J. R., Enos, F. L., Hodgson, N. C., Jacobson, W. O., Li, K. K., and Swanson, C. R., "Geothermal Loop Experimental Facility," Report No. SAN/1137-6, San Diego Gas and Electric Co., San Diego, Calif., May 1977.
[18] "Methods for Chemical Analysis of Water and Wastes," Technical Report EPA-625/6-74-003a, U.S. Environmental Protection Agency, Environmental Research Center, Cincinnati, Ohio, 1976.
[19] Fournier, R. O., *Geothermics,* Vol. 5, Nos. 1-4, 1976, pp. 41-50.

Summary

The 16 papers in this book deal with the problem of interfacial reactions between geothermal fluids and the surfaces of materials in pipes, heat exchangers, and other vessels. There are two connotations to the term "scaling" in geothermal systems. The primary meaning of the term involves the nucleation and growth of mineral scales from aqueous species of geologic origin. The term refers, secondly, to the establishment of a corrosion product deposit. Corrosion and deposition of corrosion products can promote the formation of mineral scale through roughening of the surfaces of vessel walls. The rate of scale deposition, as well as some of the physical properties of the scale (adherence, friability), is dependent on the nature of the substrate surface. As such, corrosion may have an important influence on the scaling properties of a particular system. Scaling may also inhibit or promote corrosion. In certain cases the existence of a thin, adherent scale will provide a protective coating to the underlying material. In other cases the scale may provide a pit or a crevice at the surface, initiating a localized corrosion attack.

This book is introduced by Macdonald's paper on the fundamental thermodynamics of corrosion. The paper by Conover et al provides a concise overview of the materials problems in geothermal systems. From this summary, it is evident that corrosion is complicated by the highly variable chemistry of the geothermal fluids, which are unique to a given field or even a specific well. The problem is further confounded by the existence of different modes of corrosion, such as stress corrosion cracking and pitting.

Danielson's paper on the linear polarization technique provides important information on the application of commercial corrosion monitoring instruments to geothermal systems. One of the serious corrosion problems for geothermal plants is the potential variability of the water chemistry as physical (pressure and temperature) and chemical (precipitation, gas solution/dissolution) changes occur at different points of fluid processing. Because of this, there is the need for multipoint corrosion monitoring, which *in situ* corrosion probes can provide. Schnaper et al have provided an interesting application of electrochemistry to *in situ* corrosion protection in geothermal systems.

The next six papers deal with metalic corrosion in specific geothermal systems. These range from the moderate-temperature system with low dissolved solids at Raft River, Idaho, to the high-temperature plus high dissolved solids system at the Salton Sea known geothermal resource area in California and the steam-dominated fluid at Wairakei, New Zealand.

The next four papers fall into a group concerned with the stability of polymer materials in geothermal systems. This group of papers represents nonmetallurgical approaches to the geothermal materials problems. Lorensen et al have provided a detailed study of polymers, including many composite materials that utilize glass and other inorganic materials, and of polymers coated on several metal substrates. The papers by Fontana and Zeldin and Zeldin et al consider polymer concretes as alternate materials of construction. These are composite materials with inorganic phases in a polymer matrix.

The final section of the book deals with geothermal brine chemistry. Phillips et al have provided a concise summary of state-of-the-art treatment methods for geothermal fluids for both corrosion and scale control. This approach to the problem seeks to provide protection from scaling and corrosion by altering some aspect of the brine chemistry, such as the pH, a gaseous constituent, or a dissolved substance, through physical or chemical means. The paper by J. C. Watson is concerned with sampling and chemical analysis of geothermal fluids.

L. A. Casper
T. R. Pinchback

E G & G Idaho, Inc., Idaho National Engineering Laboratory, Idaho Falls, Idaho 83401; symposium cochairmen and editors.

Index

A

Aeration (see Chemical species, Oxygen)
Alloys (see specific classes)
Aluminum, alloys, 8, 37, 58
Alterants, chemical, 214
Analysis, brine constituents, 236

B

Boron, removal, 219

C

Calcium/bicarbonate ratio, 229
Carbon and low-alloy steels, 7, 18, 30, 48
Cast iron, 86
Chain scission, 166
Chain stripping, 162
Chemical composition, brines, 73, 115, 144, 244
  Steam system, 86
  Synthetic, 45
Chemical species, 3, 73, 220
  Analysis of, 236
  Ammonia, 6, 27, 29
  Chloride, 6, 27, 29
  CO_2, 4, 6, 27, 29, 118
  Complexes, 21
  Hydrogen ion (pH), 6, 16, 26, 120
  Logging, 225
  Methane, 118
  Metal ions, 27
  Oxygen, 4, 6, 27, 29, 41, 45, 50, 114, 119, 151
  Sulfate, 6, 14, 27
  Sulfide, 6, 15, 27, 29, 155, 165
Coatings,
  On elastomers, 158, 160, 162
  On metals, 57, 170
Cobalt, alloys, 8, 36, 88
Composite materials, 29
Compressive strength, 198
Copper, alloys, 8, 36, 58, 88, 119
Corrosion, modes, 29
Crevice corrosion, 7, 27, 111, 120, 132
Cyclic voltammetry, 44, 55, 59

E

Elastomers (see Polymers)
Electrocoatings, 57
Erosion, 143
Electron spectroscopy for chemical analysis (ESCA), 63
Exfoliation, 30

F

Failure
  Elastomers, 156
  Polymeric/composite materials, 165
Field investigations, 41, 116
Fluidized bed, 69
Fouling (see Scaling)

G

Galvanic corrosion, 7, 30
Geothermal power cycles, 25
Graphitic corrosion, 103

H

Heat exchanger, tube-in-shell, 70
Heat-affected zone (HAZ), 34, 70
Hydrogen blistering, 27

I

In-plant tests, 85
Infrared spectroscopy, 178, 201
Inhibitors, 213
Intergranular corrosion, 29, 147
Iron, 86, 121
Iron, potential-pH diagram, 17

L

Lead, 88
Linear polarization, 41
Low-alloy steels, 41, 58, 70, 86, 119, 121

M

Microstructure, 147
Molybdenum, 88, 119

N

Nickel, alloys, 7, 19, 36, 88, 119
Nickel, potential-pH diagram, 16

P

Pitting, 7, 27, 55, 74, 89, 98, 103, 109, 121, 134, 148
Polarization resistance, 42
Potential-pH diagrams, 14, 48
Polymers, 39, 59, 158, 162, 168, 182
  Concrete composite, 39

R

Round-robin testing, 237

S

Scanning electron microscopy (SEM), 174, 201
Scaling, 70, 117, 165, 187, 210
  Control, 209
  Calcite, 75
  Adhesion, 166, 171
  Computer simulation
  Resistance of materials, 170, 174
Seeding, 213
Static exposure, 82
Species in solution (see Chemical species)
Stress corrosion, 7, 27, 74, 81, 124, 147
Stress–strain curves, 199
Stainless steels, 41, 70, 86, 119, 143

T

Tantalum, 37, 88
Titanium, 35, 88, 119

W

Water, meteoric and magmatic, 4
Weight loss, 41, 48, 64, 81
Weld corrosion, 129

Z

Zinc, 88
Zirconium, alloys, 8, 37, 88